REVIEW OF THE
MITE FAMILY CHEYLETIDAE

REVIEW OF THE
MITE FAMILY CHEYLETIDAE

BY

FRANCIS M. SUMMERS and DOUGLAS W. PRICE

UNIVERSITY OF CALIFORNIA PRESS
BERKELEY · LOS ANGELES · LONDON
1970

University of California Publications in Entomology
Advisory Editors: J. N. Belkin, R. M. Bohart, Paul DeBach, J. R. Douglas, R. L. Doutt,
D. D. Jensen, W. H. Lange, E. I. Schlinger
Volume 61
Approved for publication November 21, 1969
Issued September 17, 1970
Price, $5.50

University of California Press
Berkeley and Los Angeles
California

◇

University of California Press, Ltd.
London, England

ISBN: 0-520-09134-5

Library of Congress Catalog Card No.: 78-629638

CONTENTS

Contents

REVIEW OF THE
MITE FAMILY CHEYLETIDAE

BY

FRANCIS M. SUMMERS and DOUGLAS W. PRICE

INTRODUCTION

THE FAMILY Cheyletidae Leach, 1815, presently includes close to 50 genera and approximately 186 described species. The known species are sparingly distributed among the genera. At least sixteen genera are monotypic and only three have more than ten species each. The authors felt initially that the genera within the family had been too generously proliferated. Nevertheless, we have been obliged to concur with many of the taxonomic decisions of previous workers.

Although a number of species of Cheyletidae are associated with birds and mammals, most are considered to be free-living predators. They feed on a variety of other micro-arthropods, particularly on herbivorous and saprophytic forms such as acarid mites and collembolans. Cheyletids are often abundant in granaries, animal-feed warehouses, barns, and stables, especially if acarid populations are high. They are found also in leaf litter and topsoil wherever conditions support a substantial soil fauna. Some occur on the bark and foliage of fruit trees and other woody plants where they may feed on scale insects and plant-feeding mites. This project evolved out of difficulties encountered by the authors in identifying cheyletid mites taken in commonplace situations such as those noted above. Surprisingly few of the specimens collected in California would fit available generic and specific descriptions.

Opinions regarding the family affinities of the Cheyletidae differ. Cunliffe (1955) placed four families in the Cheyletoidea: Cheyletidae Leach, 1815; Myobiidae Mégnin, 1877; Demodicidae Nicolet, 1855; and Heterocheylidae Trägårdh, 1950. Dubinin (1957) grouped the Demodicidae with the Psorergatidae W. Dubinin, 1955, in the superfamily Demodicoidea; and he placed the Cheyletidae, Syringophilidae, Harpyrhynchidae, and Myobiidae in the superfamily Cheyletoidea. Southcott (1961) considered the Cheyletoidea to be comprised of the Cheyletidae, Myobiidae, Heterocheylidae, and Ophioptidae Southcott, 1956. Volgin (1966*a*) grouped the Cheyletidae, Syringophilidae, Harpyrhynchidae, and Ophioptidae in the Cheyletoidea. The close relationship between the Syringophilidae and Cheyletidae is emphasized by Volgin (1966*a*).

The basic characteristics of the Cheyletoidea are generally considered to include the presence of a palpal thumb-claw complex, the presence of peritremes, fusion of the cheliceral bases with the gnathosoma to form a rigid gnathosomal unit, stylet-like movable chelae, and homomorphous larvae. The generalized condition is best seen in the free-living Cheyletidae, with varying degrees of reduction and specialization occurring in the families noted above.

Volgin (1961) divided the family into two subfamilies of unequal size. Most of the genera known at that time were placed in the Cheyletinae Leach on the basis of possessing one or two comblike setae on the palp tarsus. Only four genera were assigned to a second subfamily, the Cheyletiellinae Volgin, 1961: *Cheyletiella*

Canestrini, *Neocheyletiella* Baker, *Hemicheyletus* Lawrence, and *Ornithocheyla* Lawrence. These genera lack comblike setae on the palp tarsus, contain only animal-associated species, and are among the more highly specialized cheyletids. On the basis of the absence of comblike setae, several other genera were recently assigned to this subfamily by Volgin (1969). These are *Bakericheyla* Volgin, *Criokeron* Volgin, *Eucheyletiella* Volgin, and *Ornithocheyletia* Volgin. However, not all animal-associated cheyletids lack the comblike setae on the palp tarsus.

Volgin's series of papers published during the period 1949 to 1966 have provided desirable and sound revisions of numerous cheyletid genera. His most recent contribution (1969) is a fine monograph on the family from the worldwide point of view. In that paper, 54 genera are sorted into ten tribal taxa, all new tribes. His reassignment of species resulted in eleven additional genera, several of which are combined in our present study. We agree heartily with many of Volgin's decisions, and the contrary opinions incorporated in our paper do not diminish our appreciation of his contributions. Two factors contribute to our divergent opinions. First, the language barrier creates some confusion about definitions and intended meanings. Second, the principal changes in cheyletid systematics introduced in this paper have resulted from new information obtained from type specimens on deposit in the United States or from specimens which we were able to borrow from abroad.

Although no distinct morphological cleavage is apparent between the free-living Cheyletidae and those associated with birds and mammals, our primary objective in this paper is to contribute to a better understanding of the free-living, predaceous forms. Also, the specimens available to us for study were primarily of this type. Therefore, with regard to the animal-associated species, we have attempted merely to assemble and to summarize the contributions of other workers.

The recognitional characters which distinguish the cheyletids from other families of predatory Prostigmata are associated with the gnathosoma and the dorsal body setae. The free segments of the pedipalps are broadly articulated with a sturdy *basis capitulum*. In life the palps are carried in a partially pendant position, curving downward as well as forward. The femur is the largest segment, being slightly elbowed in its midsection to curve and project forward. The genu and tibia are stubby and arranged to direct the prominent tibial claw mesad. The forward and downward motion of the two opposed palps produces a powerful pincer-like raptorial mechanism. Specimens mounted on microslides tend to have the palps lifted into a more prognathous position.

Although the palp tarsus is short and knoblike, its sensilla are conspicuous and unique. Five specialized sensilla are normally present: two large pectinate or comblike setae (combs), two smooth sickle-like setae (sickles), and a small inflated peglike seta or solenidion. The curvature of the combs and sickles conforms to that of the tibial claw so that, at rest, they form a bundle appressed to the mesal face of the claw. The structure of these sensilla and their relationship to each other are patent features of most cheyletids.

Another distinctive feature of the Cheyletidae is the relationship of the peritremata to the chelicerae. As noted, the basal parts of the right and left chelicerae are fused together into a broad, integral cheliceral plate or stylophore (Snodgrass, 1948). Inlaid into the surface skeleton of the stylophore are conspicuous, partially segmented peritremes. These air channels are continuous with a pair of tracheal trunks ascending in the middorsal region of the gnathosoma near the anterior border of the stylophore (T. E. Hughes, 1958). Segmented cheliceral peritremes of this type are seen also in the Syringophilidae, but are greatly reduced or absent in other Cheyletoidea.

The emergence of cheliceral peritremes distad on a stylophore occurs also in the Raphignathoid family, Caligonellidae Grandjean (Summers and Schlinger, 1955). The simpler organization of the stylophore in the Caligonellidae and the emergence of the peritremes at various locations along its middorsal line in different genera suggest that the large expanse of the cheyletid gnathosoma circumscribed by the peritremes is the basal part of the fused chelicerae. This part of the stylophore overlies the pedipalpal coxae, and seems to be integral with them. The term *tegmen* is proposed here to designate that portion of the stylophore in Cheyletidae lying proximal to the peritremes.

The term *protegmen* is proposed for that part of the stylophore occurring in front of the peritremes. The extent and the configuration of the protegmen are highly variable. Although difficult to describe, differences in the structure of the protegmen have considerable taxonomic value in distinguishing cheyletid genera.

The term *rostrum* is applied to a ventral, conical projection of the *basis capitulum*, thought to be derived from the fused endites of the pedipalps (T. E. Hughes, 1959). In most cheyletids the rostrum projects forward beneath the stylophore. The cheliceral stylets appear to rest in a groove in the dorsal surface of the rostrum. In certain genera such as *Cheletomorpha, Cheletogenes*, and *Hemicheyletia,* the protegmen is enlarged and forms an elevated bulbous or hood-like canopy overlying much of the rostrum. In these genera it appears that the protegmen articulates laterally with the sidewalls of the rostrum.

In other cheyletid genera, *Cheyletus, Eucheyletia,* and *Mexecheles,* the protegmenal part of the stylophore is somewhat conelike and drawn into a pointed apical process. This process merges with the dorsal face of the rostrum, or is tightly pressed into its dorsomedian groove. A specimen of *Cunliffella panamensis* crushed between the slide and coverglass revealed that the apical parts of the protegmen are not fused with the rostrum. In this species the apical parts are bifid, and the two projecting lobes may represent the remnants of a pair of fixed cheliceral digits.

The cheliceral stylets articulate with the protegmen by means of basal sclerites. These stylets are believed to be independently exsertile since, in slide-mounted specimens, they often protrude to varying degrees beyond the tip of the rostrum.

It is evident that the distal parts of the gnathosoma are fashioned into a fairly rigid beak, but we do not know exactly how the protegmen and rostrum are fitted together. The range of independent motion of the rostrum, in protraction and retraction, seems to be quite limited. The variation in length of the gnathosoma in a series of specimens is always very small. The piercing thrust of the mouthparts seems, therefore, to depend primarily upon the protraction of one or both of the cheliceral stylets.

The body setae encountered among different genera exhibit an extraordinary diversity of form. Thus they figure prominently in the descriptions of species and genera. Attempts by taxonomists to describe setae according to their size and qualitative differences have created a terminology using many comparative adjectives and "-like" words (e.g., staghorn-like, cloudlike, fanlike, petaloid, lanceolate). This terminology is cumbersome and inconcise, since different workers do not make the same associations of words and images. However, we have followed the conventional terminology, and hope that the illustrations will clarify our descriptions.

Another problem in cheyletid chaetotaxy is the identification of particular setae. The setae of the appendages present no special problems, and only a few new names are proposed to designate the tarsal setae: addorsals, paraterminals, infraterminals, and azygos (fig. 1). We have made no attempt to apply a rigid terminology to the dorsal body setae because in some cases their homologies remain uncertain. However, we have classed the anteriormost pair of propodosomal setae (the verticals) as dorsolaterals even though this pair often aligns with others classed as dorsomedians. Similarly, the penultimate (5th) pair of dorsolateral hysterosomal setae often is out of line with others in this longitudinal series; its two setae lie closer to the midline, and might logically be considered as a pair of dorsomedians.

The designation of these two pairs of setae as dorsolaterals is based on the fact that in the female line of many species the dorsomedian setae undergo a distinctive metamorphic change during the final moult. Whereas the dorsomedian setae of deutonymphs are orthodox and resemble the dorsolaterals, in the adult stage they are peculiarly modified and often bizarre. When these modified dorsomedian setae characterize a species, the two pairs of setae mentioned above retain the form of the dorsolaterals, and do not become modified into the form characteristic of the dorsomedians.

The mechanism controlling this differentiation of setae does not prevent considerable variation in its expression among individuals. One or another pair of dorsomedians may retain the orthodox or deutonymphal form after the final moult; or one member of a given pair in the adult may be modified while the other is not. (Rarely a dorsomedian seta may make only a partial change during the final moult (fig. 42e).

The relatively large number of genera containing only one or at most but a few species indicates that students of this group of mites are as yet unable to assess the phylogenetic implications of external characters. Thus the morphological differences which may exist among congeners are not clearly understood. In part this may be due to the fact that the known species probably represent only a small sample of the cheyletid fauna of the world.

It is difficult to capture in a few descriptive words the exact essence of generic groupings within this family. Earlier workers defined genera mostly on the basis of certain combinations of selected morphological similarities. Although numerous small differences readily serve to differentiate species within genera, discrete differences between groups of species are more difficult to recognize. For example, the form of the setae varies tremendously and provides a reliable means for separating species. It is difficult, however, to determine when the gradual or continuous differences exhibited by the setae transcend the specific realm and acquire generic importance. This holds true also for other "generic" characters such as the number and size of the dorsal plates, the nature of the tarsal claws, the number of teeth on the palp claw, the nature of the sensilla on the palp tarsus, and the structure of the peritremes.

Although there are a few species, especially in *Cheyletus*, for which only males have been described, we have dealt only with females in this study. The identification of species on the basis of male specimens presents several problems. Field samples yielding only a few individuals often do not include males. The proper matching of males with females in field collections, particularly if obtained at different times and from different localities, is uncertain. Also, the occurrence of interspecific hybrids in dense populations of mixed species (Edwards, in A. M. Hughes, 1961) and the occurrence of heteromorphic males may further complicate the identification.

All linear measurements of body parts are given in microns with the μ or *mu* symbol omitted. These measurements have limitations in respect to species discriminations, but they may give some objectivity to descriptions of size or size relationships between parts. When specimens mounted on microslides are compressed into almost flat "pelts" to spread the legs and gnathosomal appendages, the thinner membranous parts of the body may be more distorted than the more rigid sclerotized areas.

The length of the gnathosoma was measured from the tip of the rostrum to the dorsomedian base line of the stylophore. The length of the idiosoma was measured from the same base line of the stylophore to the posterior extremity of the body. The sum of these two measurements represents the total body length at the midline. The measurement for length of legs is the total length of the free movable segments only, from the coxotrochanteral articulations to the tips of the claws.

The authors have deposited type specimens in the Entomology Museum, Department of Entomology, University of California, Davis, and in the United States National Museum, Washington, D.C. The names of these depositories are abbreviated UCD and USNM, respectively.

KEY TO GENERA—FEMALES

1. All tarsi with paired claws 2
   One or more tarsi without paired claws 45
2. Eyes not evident 3
   One pair of eyes present 18
3. With 2 well-developed comblike setae on palp tarsus 4
   With 1 or no comblike setae on palp tarsus 7
4. Idiosoma fusiform, elongate; coxae of legs III and IV displaced far to rear, so that coxa
   IV encroaches upon genital covers ....*Bak* Yunker
   Idiosoma ovoid or pyriform; coxae III and IV not appreciably separated from coxae
   I and II 5
5. No major plate on hysterosoma; protegmen and rostrum almost coextensive, the latter
   barely protruding beneath the former ....*Cheletonella* Womersley
   Hysterosoma with 1 major plate; rostrum protrudes considerably beyond protegmen 6
6. Setae on margins of dorsal plates acicular, fusiform or narrow spatulate, conservatively
   barbed; no fan-shaped anal setae; dorsomedian setae, when present, few in number,
   tiny and simple in structure ....*Cheyletus* Latreille
   Setae on margins of dorsal plates broadly spatulate or fan-shaped; dorsomedian setae
   numerous, conspicuous, greatly modified ....*Eucheyletia* Baker

---

[1] The intent is to exclude from consideration the pairs of trivial platelets often associated with individual setae not implanted on the major plates.

24. Claws of tarsi II to IV with basal outgrowths ....*Neoacaropsis* Volgin
    Claws of tarsi II to IV smooth hooklets, without basal processes 25
25. Median hysterosomal plate large enough to cover most of metapodosoma and hystero-
    soma; no obvious suranal plate 26
    Median hysterosomal plate small, restricted to midportion of metapodosoma; bears no
    setae; a 2d median plate may cover opisthosoma in suranal position 27
26. Humeral seta acicular, smooth or nearly so, much longer than dorsal body setae
                                        Acaropsis Moquin-Tandon
    Humeral setae relatively short and spatulate, resembling nearby dorsal setae
                                          Acaropsella Volgin
27. Dorsal body setae acicular, smooth or nearly so; humeral setae ultralong; suranal plate
    covers tip of opisthosoma ....*Paracaropsis* Volgin
    Dorsal body setae spatulate or narrow fans; humeral setae not longer than nearest dorsal
    body setae; suranal plate not apparent ....*Cheletonata* Womersley
28. Dorsolateral body setae acicular, smooth or very lightly barbed ....*Samsinakia* Volgin
    Dorsolateral body setae not acicular, may be barbed rods or straplike blades or further
    flattened to become spatuate or flabellate 29
29. Dorsolateral setae on hysterosoma flabellate or clamshell-like; 8 or more pairs of median
    setae, some or all of these squamate and overlapped to cover most of dorsal plate 30
    Dorsolateral setae on hysterosoma rodlike, straplike, spatulate, or fanlike; 1 to 6 pairs
    of median hysterosomal setae (one exception has 9 pairs), some or all of which may
    resemble dorsolaterals or, if peculiarly modified, not as above 33
30. Gnathosoma foreshortened, with usual projection of rostrum from apex of basis capituli
    disarranged by midventral encroachment of palp femora; no subcapitular setae; peri-
    tremes form reverse loops around paired chambers in margins of stylophore
                                            Cunliffella Volgin
    Gnathosoma approximately an elongate cone with rostrum and protegmen prominently
    displayed; 1 pair of subcapitular setae; shape of peritremes normal, marginal cham-
    bers absent in stylophore 31
31. Inner sensillum, generally called the sickle-like seta of palp tarsus, distended or inflated
                                           Neoeucheyla Radford
    Inner sickle-like seta of palp tarsus acicular or conventional in form 32
32. Dorsal plates bear about 12 pairs of squamate dorsomedian setae; palp claw with teeth
    confined to its basal half ....*Cheyletia* Haller
    Dorsal plates bear about 16 to 18 pairs of squamate dorsomedian setae; palp claw with
    8 to 11 small teeth distributed along most of its length ....*Hypopicheyla* Volgin
33. Median dorsal plate absent on hysterosoma ....*Cheletacarus* Volgin
    Median dorsal plate present on hysterosoma; may be reduced in size and bear no setae 34
34. Median hysterosomal plate represented by a small setigerous sclerite or suranal plate near
    tip of opisthosoma ....*Cheletophyes* Oudemans
    Median hysterosomal plate usually completely covers hysterosoma, or may be reduced to
    a small setigerous sclerite over metapodosoma 35
35. Marginal setae of propodosoma heteromorphic, those in front of eyes broad flabellae,
    those behind eyes long, narrow, straplike ....*Grallacheles* De Leon
    Marginal setae of propodosoma similar to each other in size and shape 36
36. Venter of opisthosoma with superfluous setae: 4 pairs resembling dorsal setae arranged
    in a crescentric row across anal area, these in addition to 2 or 3 pairs of short anals
                                          Cheletophanes Oudemans
    Venter of opisthosoma without superfluous setae; anal papilla with 2 or 3 pairs of short
    setae, one of which may be fan-shaped 37
37. Leg I equal to or longer than idiosoma (see page 5 for definition of measurements) 38
    Leg I shorter than idiosoma 39
38. Palp claw with 7 to 12 teeth; 2 setae on femur IV ....*Mexecheles* De Leon
    Palp claw with 1 tooth; 1 seta on femur IV ....Cheletomorpha Ouds. (in part)

Cheletomimus Oudemans

Cheletomimus Oudemans, 1904*b*, Ent. Ber. Nederl. Ent. Ver. 1(18) :163.

One pair of plates partially covers metapodosomal area of adult female. Median dorsal setae of idiosoma similar to laterals in form and size. Eyes present.

Type species.—Cheletes berlesei Oudemans, 1904*a*, Ent. Ber. Nederl. Ent. Ver. 1(17) :154. Monotypic.

Three species were originally assigned to this genus. Of the two more recently described, *duosetosus* Muma appears to be more closely related to the type species

---

[2] *Hoffmannita clavipes* (Volgin), 1963, is described as having no eyes. The type species *mexicana* Pelaez, possesses eyes.

than is *denmarki* Yunker. Except for the nymph-like arrangement of the hysterosomal plates, *berlesei* and *duosetosus* closely resemble the less ornate species of *Hemicheyletia*. The affinities of *denmarki* are less obvious, and this species was recently removed to a separate genus, *Oudemansicheyla*, by Volgin (1969).

KEY TO FEMALES OF CHELETOMIMUS

1. With 3 pairs of median setae on propodosomal plate; 1 seta on genu IV . . . .*berlesei* Oudemans
   With 1 pair of median setae on propodosomal plate; 2 setae on genu IV . . . . .*duosetosus* Muma

Cheletomimus berlesei (Oudemans)

(Fig. 2)

Cheletes berlesei Oudemans, 1904*a*, Ent. Ber. Nederl. Ent. Ver. 1(17):154; new name for species
    which Berlese (1886) misidentified as *Cheyletus ornatus* Can. et Fanz., 1876.
Cheletomimus trux Oudemans, 1904*b*, Ent. Ber. Nederl. Ent. Ver. 1(18):163.
Cheletomimus ornatus (Berl.), Oudemans, 1906, Mém. Soc. Zool. France 19:136–139.
Cheletomimus berlesei (Oudemans), Baker, 1949, Proc. U.S. Nat. Mus. 99(3238):293–294;
    Volgin, 1955, Akad. Nauk S.S.S.R., Zool. Inst., Opredel. p. Faune S.S.S.R. no. 59:171–172;
    Yunker, 1961, Canadian Ent. 93(11):1032; De Leon, 1962, Florida Ent. 45(3):134.

Palp claw bears 6 or 7 teeth; 5th or 6th tooth more robust than others. Outer comb 16 teeth; ends in a long, flat, hooked blade; inner comb ca. 25 teeth. Sickle-like setae normal. Dorsal palptibial seta lanceolate, flattened. Dorsal setae on palp femur and genu fan-shaped, much broader than others of body or appendages. Upper cheek of palp femur denticulate. Protegmen elevated, bluntly rounded, slightly concave across front margin; about 0.4 times as long as tegmen; closely set rows of bacillus-like tubercles cover both tegmen and protegmen, rows lengthwise in front of peritremes, predominantly transverse on tegmen. Rostrum wide, rounded apically, convexly curved at sides; inferior adoral setae (38) more than twice as long as superior adorals (16); about two-thirds of rostrum projects beyond protegmen. Peritremes form an inverted U, each arm with 7 links.

Propodosomal plate quadrangular, incompletely covering tergum, decorated with intricate scrollwork of tuberculate striae. Hysterosomal plating restricted to one pair of rounded sclerites on metapodosoma, each plate larger in diameter than length of single seta it bears; all other hysterosomal setae arise on minute, trivial platelets. Interscutal integument with 2-lined striae having faint tubercles. Eyes large, with convex corneas, encircled by 4 to 6 concentric striae. Dorsal body setae fairly short (20–35), lanceolate, fluted but not conspicuously barbed; 14 pairs plus humerals, all similar in form. Dorsomedian setae: 3 pairs on propodosomal plate, 1 pair on hysterosoma, near humeral sulcus.

Length ratio: leg I/idiosoma = 0.6. Setae on legs I to IV: femora 2-2-2-1, genua 3-2-2-1, tibiae 5-4-4-4, tarsi 9-8-7-7; single seta on genu IV may be unique for this species. Tarsus I: solenidion wI long (34), with minute guard seta close by its base, both on a prominent nipple. Measurements (n = 10): length idiosoma 296±16, gnathosoma 111±4, leg I 183±8, tarsus I 78±2.6.

Collection data.—Bark of willow twigs, Solano County, Calif. (S. F. Bailey); eucalyptus bark, Burlingame, Calif. (D. W. Price); soil in Municipal Rose Garden, Oakland, Calif. (R. O. Schuster); on dwarf apples, Auckland, New Zealand (E. Collyer).

Cheletomimus duosetosus Muma

(Fig. 3)

Cheletomimus duosetosus Muma, 1964, Florida Ent. 47(4):242–243.

This species is very much like *berlesei*, as indicated by the illustrations. There are two unmistakable differences: *duosetosus* has one pair of median setae on its propodosomal plate and two setae on genu IV; *berlesei* has three pairs of median propodosomal setae and only one seta on genu IV. There are at least four other, less obvious morphological differences between these two species. In *duosetosus* the microtubercles on the tegmen are circular to ovoid, tightly packed together, not obviously aligned in rows; the 6 to 8 teeth on the palp claw are almost evenly graduated in size between basal and distal members; the setae of the fifth pair of marginal hysterosomals are closer to those of the sixth (terminal) pair than to the fourth pair; the dorsal body setae are quite small (16–23). In *berlesei* the microtubercles of the tegmen are longer than wide and aligned into discernible transverse rows; the penultimate tooth of the palp claw is larger than either of the two adjacent teeth; the setae of the fifth pair of marginal hysterosomals are internal to and nearly aligned with the setae of the fourth pair to form a crossrow; the dorsal setae are larger (20–35) than those of *duosetosus*.

Collection data.—From citrus litter, Weirsdale, and sand pine litter, Lake Placid, Florida. Specimens have been checked with USNM holotype.

Oudemansicheyla Volgin

Oudemansicheyla Volgin, 1969, Akad. Nauk S.S.S.R., Zool. Inst., Opredel. p. Faune S.S.S.R. no. 101:272.

This genus was proposed to accommodate the species heretofore called *Cheletomimus denmarki* Yunker; which was designated by Volgin to be the type of the genus. *O. denmarki* differs appreciably from *Cheletomimus berlesei,* and this split-up of *Cheletomimus* produces two fairly well-defined genera.

Oudemansicheyla differs from *Cheletomimus* in these respects: the palp tibial claw bears teeth along its entire concave margin; dorsal setae are uniformly rounded or clamshell in form; the propodosomal plate bears six pairs of dorsolateral setae; dorsomedian setae are numerous on the propodosoma and the hysterosoma; the dorsal metapodosomal plates are paired, as in *Cheletomimus,* but each plate bears seven setae.

Oudemansicheyla denmarki (Yunker)

(Fig. 4)

Cheletomimus denmarki Yunker, 1961, Canadian Ent. 93(11):1032–1035; Muma, 1964, Florida Ent. 47(4):244.
Oudemansicheyla denmarki (Yunker), Volgin, 1969, Akad. Nauk S.S.S.R., Zool. Inst., Opredel. p. Faune S.S.S.R. no. 101:273.

Individuals small in size. Palp claw partly enveloped by 2 leaflike setae of palp tibia, toothed on entire margin, ca. 15 teeth. Outer and inner combs with ca. 16 teeth each. All setae of palp femur and genu leaflike. Rostrum bears a pair of anteriorly directed, nipple-like processes on which superior adoral setae arise. Protegmen elevated, domed, with thick varicose striae disposed as illustrated. Tegmen wider than long, flared outward over each maxillicoxa; surface pebbled with very coarse bacillus-like tubercles. Peritremes with 5 links per side, without abrupt curvature rearward, end link distended, positioned beneath lateral overhang of tegmen. Eyes present. Dorsal body plating liberally pebbled with tubercles, some forming irregularly swollen rugae. Interscutal integument of dorsum with strong, widely spaced tuberculate striae. Dorsal plates include an extensive propodosomal shield and a pair of hysterosomal shields, each hysterosomal shield large enough to carry 7 setae. Dorsal body setae of clamshell type, numerous, 26 pairs plus humerals, all similar in form, relatively small (12–23); 6 pairs in dorsolateral row on propodosomal plate; each seta has a thick blade, with 5 or 6 strongly raised, barbed ribs on convex face and about 12 thick radial ribs on concave (inferior) side; radial ribs overreach periphery to project as marginal spikelets. With 1 pair of anal setae clamshell-like, 2 pairs acicular, smooth or nearly so.

Length ratio: leg I/idiosoma = 0.6 Setae on legs I to IV: femora 2-2-2-1, genua 3-2-2-2, tibiae 5-4-4-4, tarsi 10-8-7-7. Tarsis I bears a leaflike guard seta which forms a loose sheath over body of tarsus; solenidion wI slender, about as long as greatest diameter of tarsus, set considerably in front of guard seta. Measurements (n = 2): length idiosoma 205, gnathosoma 70, leg I 127, tarsus I 51.

Collection data.—From leaf litter of citrus and palmetto, from several localities in central Florida.

The examples studied were collected expressly for the authors by Dr. M. H. Muma.

Hemicheyletia Volgin

Hemicheyletia Volgin, 1969, Akad. Nauk S.S.S.R., Zool. Inst., Opredel. p. Faune S.S.S.R. no. 101:201–202.
Dendrocheyla Volgin, 1969. New synonym.
Andrecheyla Volgin, 1969. New synonym.

Palp claw bears 6 to 11 teeth. Stylophore extends considerably in front of peritremes as elevated protegmen. Rostrum protrudes well beyond front fold of protegmen. Peritremes approximately horseshoe-shaped, rearward bend not acute, 5 or 6 links per side, typically 3 links in descending arms; no part of transverse arms anterior to median point of origin. Eyes present, usually large, protruding. Hysterosomal plate highly variable, usually covering most of hysterosoma but may be reduced and confined to tergal area of metapodosoma. Dorsolateral body setae relatively short, narrow spatulate to fan-shaped, none appreciably longer than others; 4 pairs on propodosomal plate, 0–5 pairs on median hysterosomal plate. Humeral setae resemble dorsolaterals, never acicular or flagelliform. Dorsomedian setae may be conventional or aberrant, with one or more pairs fragmented or staghorn-like. Solenidion wI conspicuously long, with or without a much shorter guard seta in duplex position. Tarsi I to IV with claws and rayed empodia.

Type species.—*Paracheyletia bakeri* Ehara; designated by Volgin, 1969.

The combining of Volgin's three genera simplifies the systematics of this group of species. The creation of this taxon has the effect of removing from *Parachey-letia* an assemblage of species which exhibit the *bakeri-wellsi* type of organization. After examining the types of all but three of these species (*anarbora* De Leon, *wellsina* De Leon, *bregetovae* Volgin), it is our conclusion that no special purpose is served by separating them into several genera according to whether the dorsomedian setae are fragmented or spatulate, or whether the hysterosomal plate is extensive or reduced. Species having orthodox dorsomedians are grouped with species having aberrant dorsomedians in other cheyletid genera, for example in *Paracheyletia* and *Prosocheyla*. Furthermore, the general resemblance of these species to one another is fairly clear except for *rostella* n. sp. a transitional form.

The type species, *Hemicheyletia bakeri*, has conventional dorsomedian setae which may represent the generalized condition, whereas the occurrence of aberrant dorsomedian setae is a derived or more specialized condition.

KEY TO FEMALES OF HEMICHEYLETIA

1. Dorsomedian body setae conventional in form, i.e., similar to dorsolaterals 2
   Dorsomedian setae aberrant, i.e., unlike dorsolaterals 7
2. Hysterosomal plate reduced to a small area on metapodosoma; bears only 1 pair of
   setae ....*scutellata* (De Leon)
   Hysterosomal plate bears at least 3 pairs of dorsolateral setae 3

3. With 1st pair of dorsolateral hysterosomal setae on principal plate 4
   With 1st pair of dorsolateral hysterosomal setae on separate, trivial platelets 5
4. With 1 pair of dorsomedian propodosomal setae ....*cordovensis* (De Leon)
   With 3 pairs of dorsomedian propodosomal setae ....*granula*, n. sp.
5. With 3 pairs of dorsomedian setae on hysterosomal plate ....*rostella*, n. sp.
   With 1 pair of dorsomedian setae on hysterosomal plate 6
6. With 3 pairs of dorsomedian setae on propodosomal plate; tegmen densely stippled with
   spheroid microtubercles ....*congensis* (Cunliffe)
   With 2 pairs of dorsomedian setae on propodosomal plate; tegmen with thickened
   segments of broken striae, chiefly longitudinal ....*bakeri* (Ehara)
7. Hysterosomal plate reduced to a small area on metapodosoma; bears 0–2 pairs of dorsolateral setae 8
   Hysterosomal plate with at least 3 pairs of dorsolateral setae 9
8. Hysterosomal plate bears 2 pairs of dorsolateral setae plus 1 pair of dorsomedian setae
   serrula, n. sp.
   Hysterosomal plate almost obsolete, identifiable only as a small patch of peculiarly
   whorled striae; this carries no setae ....*volgini* (Cunliffe)
9. With 2 pairs of dorsomedian setae on hysterosomal plate ....*wellsi* (Baker)
   With 1 pair of dorsomedian setae on hysterosomal plate 10
10. With 1 pair of dorsomedian setae on propodosomal plate ....*anarbora* (De Leon)
   With 2 or 3 pairs of dorsomedian setae on propodosomal plate 11
11. Dorsomedian setae minute, amoeboid rosettes with about seven conical lobules (fig. 11*g*) ;
   dorsal setae of palp tibia flattened into narrow blade fringed with barbs ...*darwinia*, n. sp.
   Dorsomedian setae staghorn-like or fragmented (as in *wellsi*, fig. 10*b*) ; dorsal setae of
   palp tibia acicular, smooth or nearly so 12
12. With 2 pairs of dorsomedian setae on propodosomal plate ....*wellsina* (De Leon)
   With 3 pairs of dorsomedian setae on propodosomal plate ....*bregetovae* (Volgin)

Hemicheyletia bakeri (Ehara)
(Fig. 5)

Paracheyletia bakeri Ehara, 1962, Ann. Zool. Japon. 35(2):109–111; Muma, 1964, Florida Ent.
   47(4):245.
Hemicheyletia bakeri (Ehara), Volgin, 1969, Akad. Nauk S.S.S.R., Zool. Inst., Opredel. p.
   Faune S.S.S.R. no. 101:202–203.

Palp claw with 7 basal teeth. Outer comb on palp tarsus ca. 16 teeth; inner comb ca. 20 teeth.
Dorsal seta on palp femur a broad, truncate fan. Tegmen with closely packed, thickened segments
of broken striae, principally longitudinal. Protegmen similarly adorned. Rostrum broad, spoon-
shaped distally, with prominent marginal lamellae supporting superior adoral setae. Dorsal
plating essentially microtuberculate; tubercles lozenge-shaped swellings disposed in sinuous rows;
tuberculate pattern grades into broken striae close to plate margins; rows converge or whorl at
several points on each plate. Eyes comparatively large, protuberant, encircled by 5 or 6 concentric rows of unbroken striae. Dorsal setae: 13 pairs plus 1 pair of humerals, all small (20–27),
broadly spatulate or cuneate, rounded at ends, fluted, 4 to 6 barbed ribs on convex face; blade
part of each seta arises from short but characteristic stalk. First pair of dorsolateral hysterosomals set on small platelets, 4 pairs on outer margins of main hysterosomal plate. Dorsomedians:
3 pairs, not structurally distinguishable from dorsolaterals, 2 pairs on propodosomal plate, 1
pair on hysterosomal plate. Integumental striae plain or faintly granular.

Length ratio: leg I/idiosoma = 0.7. Distribution of setae on two distal segments of legs I to
IV: tibiae 5-5-4-4; tarsi 9-8-7-7. Tarsus I: solenidion *w*I long (34), set on unusually high nipple;
guard seta diminutive usually not discernible; ventral seta *v* smooth. Measurements (n = 5):
length idiosoma 275, gnathosoma 98, leg I 187, tarsus I 81.

Collection data.—Leaf mold, Encinitas, Calif.; Bermuda grass (*Cynodon dactylon*), Arica,
Tarapaca, Chile (taken from grass infested with Bermuda grass stunt mite, *Aceria cynodonis*
or *A. neocynodonis*) (R. H. Gonzalez).

Hemicheyletia granula, new species

(Fig. 6)

Stylophore and rostrum together form an almost straight-sided cone with rounded apex, slightly incurved where peritremes turn rearward. Palp claw with 4 to 6 basal teeth. Outer comb ca. 15 teeth; inner comb > 20 teeth. Dorsal seta on palp femur a compact bushy plume, not shaped like dorsolateral setae. Body plates, stylophore, and proximal leg segments bear numerous rounded microtubercles; those on legs disposed in rows, on striae; those on tegmen and dorsal plates more or less evenly dispersed stipples showing little tendency to align in rows. Protegmen covers proximal two-thirds of rostrum, its anteriormost fold crescentic, thin-walled, difficult to observe. Peritremes horseshoe-shaped, 6 or 7 links each side, without abrupt changes in curvature; descending arms embedded in vertical sidewalls of tegmen. Eyes inconspicuous, corneas slightly convex, not encircled by concentric striae. Dorsal body setae short (18–25), broadly spatulate, rounded on ends, fluted, 4 or 5 branching ribs coarsely serrate, with thickened, blunt barbs; blade part of each seta begins to flare close to its basal origin. Dorsal setae: 16 pairs, 4 pairs of dorsolaterals and 3 pairs of dorsomedians on prodosomal plate, 3 pairs of dorsomedians on hysterosomal plate; 1st dorsolateral hysterosomal setae may be set on small isolated platelets or on anterior margin of main plate. Integumental striae granular except intercoxal (sternal) striae smooth.

Length ratio: leg I/idiosoma = 0.6. Distribution of setae on two distal segments of legs I–IV: tibiae 5-4-4-4, tarsi 9-8-7-7. Tarsus I: solenidion wI long (29), guard seta about one-third as long, smooth; ventral seta v barbed. Measurements (n = 8): length idiosoma 290, gnathosoma 99, leg I 184, tarsus I 77.

Types.—Holotype, female, from moss and ant nest, 1 km N, 1 km E of Quezaltepeque, El Salvador (M. E. Irwin); deposited in UCD. Eight paratypes, same sample; one deposited in USNM.

This new species closely resembles *bakeri* and *congensis*. The most obvious distinctions between the three species are evident in plate ornamentation and numbers of dorsomedian setae.

Hemicheyletia congensis (Cunliffe)

(Fig. 7a)

Paracheyletia congensis Cunliffe, 1962, Proc. Ent. Soc. Wash. 64(3):197.
Hemicheyletia congensis (Cunliffe), Volgin, 1969, Akad. Nauk S.S.S.R., Zool. Inst., Opredel. p. Faune S.S.S.R. no. 101:203–204.

Palp claw bears 7 to 11 teeth. Rostrum spade-shaped, somewhat constricted behind superior adoral setae. Tegmen densely stippled with microtubercles a few of which are noticeably larger than the more abundant smaller ones; tubercles separated from one another by interspaces less than their diameters; tubercles become rodlets close to peritremes and basal area of muscle attachments. Peritremes with 6 links per side, 1st pair comma-shaped, curved rearward as well as outward behind median origins. Dorsal plating sparsely stippled with tubercles slightly larger than those on tegmen, with little emphasis on rowlike organization; hysterosomal plate shaped like inverted triangle rounded at posterior apex. Interscutal striae sparsely tuberculate. Dorsal body setae very short (16–32), spatulate or narrow, triangular fans, largest with 3 to 5 ribs on upper face; 14 pairs plus humerals, all similar in form. With 4 pairs of marginals and 3 pairs of medians on propodosomal plate; 4 pairs of marginals and 1 pair of medians on hysterosomal plate; 1st and 6th marginals on hysterosoma set on independent platelets. Tibia I with 5 setae; a tiny spinelike guard seta in duplex position with wI.

Collection data.—Illustrated specimen, one of several females on type slide, USNM 2880, taken from *Acacia*, Kysenyi, Belgian Congo (E. W. Baker).

H. congensis belongs to the *bakeri* group. Its idiosomal plating is sparsely granulate, as in *granula*, but the ornamentation of the tegmen resembles *volgini* and *serrula*.

Hemicheyletia volgini (Cunliffe)

(Fig. 7*b*)

Paracheyletia volgini Cunliffe, 1962, Proc. Ent. Soc. Wash. 64(3):197–198.
Dendrocheyla volgini (Cunliffe), Volgin, 1969, Acad. Nauk S.S.S.R., Zool. Inst., Opredel. p.
 Faune S.S.S.R. no. 101:210–211.
Hemicheyletia volgini (Cunliffe). New combination.

This species belongs to the *wellsi* group and most nearly agrees with the description given here for *serrula*. There are two distinctive features: tibia ɪ has six setae —the additional one is a spatulate dorsomedian; the plated part of the hysterosoma is represented by a small patch of broken striae between the second and third pairs of dorsolateral hysterosomals. These broken striae course longitudinally, counter to the direction of the surrounding striae on the dorsal cuticle. None of the hysterosomal setae arise on this so-called plate.

Collection data.—Illustrated specimen, holotype USNM 2881, taken from Australian pine, INEAC, Mulunga, Belgian Congo (E. W. Baker).

Hemicheyletia cordovensis (De Leon)

(Fig. 8)

Cheyletia cordovensis De Leon, 1962, Florida Ent. 45(3):129–130.
Hemicheyletia cordovensis (De Leon), Volgin, 1969, Akad. Nauk S.S.S.R., Zool. Inst., Opredel.
 p. Faune S.S.S.R. no. 101:204–205.

The holotype female is so badly deteriorated that only a few useful details can be seen. No paratypes exist. It is unmistakably a species of *Hemicheyletia* and, if the setation is correctly interpreted, the species belongs to the *bakeri* group.

The species appears to be identifiable according to three characters. (1) The palp claw bears eight teeth, no one of which is appreciably larger than the others. (2) The dorsomedian and dorsolateral setae are similar in form. There are only two pairs of dorsomedians, one pair on each of the two major shields. (3) The dorsal plating is ornamented with tuberculate striae. The tubercles are relatively large and sparsely distributed on the idiosomal plates. The tubercles on the tegmen are equivalent in size to those on the body, but there are more of them per unit area.

Hemicheyletia scutellata (De Leon)

(Fig. 9)

Cheyletia scutellata De Leon, 1962, Florida Ent. 45(3):130.
Paracheyletia scutellata (De Leon), Muma, 1964, Florida Ent. 47(4):246.
Andrecheyla scutellata (De Leon), Volgin, 1969, Akad. Nauk S.S.S.R., Zool. Inst., Opredel.
 p. Faune S.S.S.R. no. 101:219–220.
Hemicheyletia scutellata (De Leon). New combination.

Palp claw bears 5 to 7 teeth. Outer comb ca. 16 teeth; inner comb ca. 22 teeth. Dorsal palptibial seta flattened or lanceolate, smooth. Protegmen truncate or slightly concave on front margin, about as long as tegmen; upper surfaces of tegmen and protegmen covered with modified striae; striae broken into rows of lozenge-shaped tubercles, rows generally lengthwise in direction but swirled into transversely oriented rows on posterior part of tegmen. Peritremes with 6 links per side. Dorsal plating feebly sclerotized, stippled with coarse tubercles about as illustrated. Dorsal interscutal membrane with tuberculate striae; tubercles faint and smaller than those on shield areas. Anterior shield covers most of propodosoma; posterior shield small, covering only a part of metapodosoma and bearing only 1 pair of setae. Dorsal body setae short (14–27), narrow spatulate or lanceolate; 12 pairs plus humerals. Dorsomedians and dorsolaterals alike in form; 2 pairs of medians, 1 pair of propodosomal plate, 1 pair on interscutal membrane in front of hysterosomal plate.

Length ratio: leg I/idiosoma = 0.6. Distribution of setae on distal leg segments: tibiae 5-4-4-4, tarsi 9-8-7-7. Solenidion wI (41) with very short, inconspicuous guard seta. Measurements (n = 2): length idiosoma 266, gnathosoma 105, leg I 162, tarsus I 66.

Collection data.—Illustrated paratype specimens taken from *Sweitenia mahagoni*, Coral Gables, Florida (D. De Leon); and *Xylosoma elliptica*, Tuxtla Gutierrez, Chiapas, Mexico (D. De Leon).

The small, narrow blades of the dorsal body setae and the small area covered by the hysterosomal plate are diagnostic.

<h2 style="text-align:center">Hemicheyletia wellsi (Baker)</h2>

(Fig. 10)

Cheyletia wellsi Baker, 1949, Proc. U.S. Nat. Mus. 99(3238):300–301.

Paracheyletia wellsi (Baker), Volgin, 1955, Akad. Nauk S.S.S.R., Zool. Inst., Opredel. p. Faune S.S.S.R. no. 59:169; Muma, 1964, Florida Ent. 47(4):245–246.

Cheyletia wellsi Baker, De Leon, 1962, Florida Ent. 45(3):132. Male.

Dendrocheyla wellsi (Baker), Volgin, 1969, Akad. Nauk S.S.S.R., Zool. Inst., Opredel. p. Faune S.S.S.R. no. 101:211–213.

Hemicheyletia wellsi (Baker). New combination.

Palp claw bears 6 to 9 teeth. Outer comb ca. 15 teeth, inner comb > 25 teeth. Tegmen covered with microtubercles of two sizes, smaller outnumbering larger by about 5 to 1; transverse broken striae cover basal area of muscle attachment. Protegmen broader than rostrum, vaulted, truncate in front; cuticula with a few tubercles larger than any on tegmen and equal to coarsest tuberculi on upper aspect of palp femur. Rostrum broad, spoon-shaped, with marginal lamellae supporting superior adoral setae. Dorsal plating on idiosoma microtuberculate, tubercles preponderently of one size; a few on central area of propodosoma becoming elongate and forming several whorls of "dotted" striae; striae nearest midline of propodosomal plate serpentine, more transverse than longitudinal in direction. Eyes prominent, corneas large (12), protuberant, each encircled by 5 or 6 rows of concentric striae. Each peritreme with its 1st link curving outward behind its point of origin, 6 links each side; links in descending arm individually curved and joined into a chain. Dorsal setae: 16 or 17 pairs plus humerals. Dorsolateral setae spatulate, almost straight-sided blades without peduncles; those on propodosoma appreciably wider than those on opisthosoma, ends gently rounded or nearly truncate, 2 to 4 barbed ribs. With 1st pair dorsolateral hysterosomals on tiny individual platelets; subsequent 4 pairs on margins of hysterosomal plate. Dorsomedian setae highly modified, "staghorn-like" or otherwise describable as clusters of irregularly shaped sclerotic particles strung together on radial branches of hyaline matrix; 4 or 5 pairs on propodosomal plate, 2 pairs on anterior half of hysterosomal plate. Integumental striations appear as double folds with irregularly spaced tubercles; ventral body striae without tubercules.

Length ratio: leg I/idiosoma = 0.7. Distribution of setae on two distal segments of legs I to IV: tibiae 5-5-4-4, tarsi 9-8-7-7 (guard seta counted). Tarsus I: solenidion wI long (38), guard seta extremely small, obscure; ventral seta v sparsely barbed. Measurements (n = 3): length idiosoma 302, gnathosoma 105, leg I 201, tarsus I 83.

Collection data.—Soil sample, Buenos Aires, Argentina; under *Opuntia*, La Cruz, Chile; from grape leaves, La Cruz, Chile; from Bermuda grass (*Cynodon dactylon*), Hotel El Morro, Arica, Tarapaca, Chile (R. H. Gonzalez).

<h2 style="text-align:center">Hemicheyletia serrula, new species</h2>

(Fig. 11a–f)

Palp claw with 6 or 7 teeth. Outer comb ca. 18 teeth; inner comb ca. 25 teeth. Tegmen bears countless microtubercles, some smaller than others, but one size blends with other sizes in a graduated series; tuberculate organization changes into broken longitudinal striae over basal area of muscle attachment. Protegmen broader than rostrum, covered with ovoid microtubercles, most larger than on tegmen. Tip of rostrum between superior adoral setae fashioned into a cone, not hyperbolic or spoon-shaped. Links of peritremes not individually bowed or curved except to conform with general curvature of chain of links in each arm; 3d, 4th, and 5th links noticeably inflated or distended. Dorsal plating microtuberculate; tubercles subcircular near margins of plates, gradually transforming into short tapered rods on plate midsections; tubercles and rods disposed in rows which reveal their origins as nodules on striae; these modified striae converge and whorl in complex configurations around or between setae; direction of striae principally longitudinal on middle part of propodosomal plate. Rearward extent of hysterosomal plate indeterminate, blending into less sclerotized cuticle close behind 3d pair of dorsolateral hysterosomal setae; striae of generalized integument double-lined, tuberculate on dorsum, plain and smooth on venter. Eyes as large as in *wellsi*, with concentric striae. Dorsal setae: 14 pairs (23–29) plus humerals; 1st dorsolateral hysterosomals independent of median plate; 2d and 3d pairs on hysterosomal plate; 4th pair in transitional zone, probably not supported by this plate. Dorsomedian setae aberrant; each has a stubbed shaft barely projecting above rim of its alveolus and a compact cluster of 12 to 15 indistinct, flocculose "sclerites"; 3 pairs on propodosomal plate, 1 pair on hysterosomal plate.

Length ratio: leg I/idiosoma = 0.8. Distribution of setae on two distal segments of legs I to IV: tibiae 5-5-4-4, tarsi 8-8-7-7, Tarsus I: solenidion wI very long (57), guard seta not observed; ventral seta v with 2 or 3 fine barbs. Measurements (n = 2): length idiosoma 312, gnathosoma 127, leg I 242, tarsus I 101.

Types.—Holotype, female, from rotting vegetation in rock crevice, Darwin Research Station, Galapagos Islands (R. O. Schuster); deposited in UCD. One paratype, female, from roots of grass, Darwin Research Station, Galapagos Islands (R. O. Schuster); deposited in USNM.

There are a number of subtle differences between *wellsi*, *serrula*, and *darwinia*. The cuticular ornamentation on the tegmen of *serrula* closely resembles that of *darwinia* (figs. 11*f*, 11*h*). Both are distinguishable from that of *wellsi* (fig. 10*a*). The transparent area of muscle attachment across the base of the tegmen in *wellsi* has predominantly transverse striae, whereas in *serrula* and *darwinia* these striae are predominantly longitudinal. Several links in the peritremes of *serrula* are distended or inflated, and none are markedly incurved to produce the "scalloped" effect in the series of links as seen in *darwinia* and *wellsi*. The dorsomedian setae in *serrula* (fig. 11*b*) are less elaborately fragmented than in *wellsi* (fig. 10*b*). The dorsomedian setae of *darwinia* are more rosette-like in shape (fig. 11*g*). The occurrence of one pair of dorsomedian setae on a shortened hysterosomal plate is the general tendency in *serrula* and *darwinia*, whereas two pairs of dorsomedian setae occur on this plate in *wellsi*.

Hemicheyletia darwinia, new species

(Fig. 11*g, h*)

Palp claw with 8 teeth. Outer comb ca. 17 teeth; inner comb ca. 25 teeth. Dorsal seta of palp tibia flattened into a narrow blade fringed with barbs (this seta flattened but not barbed in *wellsi, serrula*). Cuticula of tegmen, protegmen, and upper cheeks of palp femora pebbled, with relatively coarse (2–3μ), circular to oval tubercles; size range of tubercles similar on each of above-named parts; basal section of tegmen with faint longitudinal striae. Dorsal body plating also with round to oval tubercules like those on mouthparts, but somewhat less crowded together and more distinctly aligned in rows; rows on propodosomal plate between dorsomedian setae somewhat serpentine yet with tendency to run longitudinally, definitely not transversely. Eyes as in *wellsi*. Dorsal setae: 10 pairs of dorsolaterals (20–23), 3 pairs of dorsomedians; form of dorsolaterals as illustrated for *serrula* (fig. 11*d*); 1st pair of dorsolateral hysterosomals on independent platelets; hysterosomal plate extends rearward to 5th pair of dorsolaterals, i.e. 4 pairs originate on margins of this plate. Dorsomedian setae aberrant; each seta comprises a minute stalk tipped with rosette of 5 to 7 inflated conical lobules each having a thickened or sclerotized end-cap; 2 pairs on midsection of propodosomal plate, 1 pair on anteriormost margin of hysterosomal plate. General body integument with double-lined tuberculate striae covering dorsum, smooth striae on venter.

Length ratio: leg I/idiosoma = 0.6. Distribution of setae on two distal segments of legs I to IV: tibiae 5-5-4-4, tarsi 8-8-7-7 (guard seta not counted). Tarsus I: solenidion wI long (39); guard seta apparently absent; ventral seta v fringed with 5 or 6 barbs. Measurements (holotype): length idiosoma 296, gnathosoma 105, leg I 187, tarsus I 74.

Holotype.—Female, from grass near bay shore, Darwin Research Station, Galapagos Islands (R. O. Schuster); deposited in UCD.

This species is a very close relative of *serrula*. The number and character of the dorsomedian setae are possibly differentiating spot characters. *H. darwinia* has a barbed seta on the dorsal face of the palp tibia, a scalloped effect in the linkage of peritreme segments (as in *wellsi*), comparatively coarse tubercles over most of the stylophore, and a hysterosomal plate long enough to accommodate four pairs of dorsolateral setae.

Hemicheyletia rostella, new species
(Fig. 12)

Organization of palps as described for *Paracheyletia pyriformis*. Palp claw with 11 or 12 teeth. Stylophore formed into elongate cone slightly constricted where peritremes turn rearward. Protegmen truncate or slightly concave in front; cuticula with coarse tubercles, irregularly scattered, interspersed with short segments of very faint varicose striae; tubercles subequal in size on all parts of axial skeleton. Exposed parts of rostrum slender, tapered to blunt point; relative lengths of tegmen, protegmen, and rostrum = 16:9:9, respectively. Eyes prominent, protruding, encircled by 1 or 2 rows of tubercles. Hysterosomal plate narrower than propodosomal plate and separated from it by a wide band of striae, covering most of hysterosoma; bears 4 pairs of dorsolateral setae. Integumental striae tuberculate or varicose on dorsum, smooth on venter. Dorsal body setae: 14 to 16 pairs (23–29) plus humerals, all structurally alike, distinctively sculptured; each seta a broad, transparent spatula with radial ribs of underside sharply outlined and projecting as evenly spaced spikelets from rounded end of blade; uppermost ribs represented by 6 or 7 rows of uniformly spaced barbs; each barb appears as an isolated or discrete sclerotic spine. With 1st pair of dorsolateral hysterosomal setae on small platelets remote from principal plate. Dorsomedian setae: 2 or 3 pairs on propodosomal plate, 2 or 3 pairs on hysterosomal plate.

Length ratio: leg I/ idiosoma = 0.8. Distribution of leg setae: tibia 5-5-4-4, tarsi 9-8-7-7. Tarsus I: pedicel slender but not abnormally shortened; claws diminutive (ca. 5μ), about half as long as claws on tarsus II; paraterminal setae not excessively lengthened; solenidion wI very long (42); guard seta less than 0.2 times as long as wI; ventral seta v barbed. Measurements (n = 10): length idiosoma 342 ± 15, gnathosoma 137 ± 5, leg I 262 ± 16, tarsus I 98 ± 5.

Types.—Holotype, female, from rotting logs, Post Pile Camp, Tehama Co., Calif. (R. O. Schuster and A. A. Grigarick); deposited in UCD. Numerous paratypes from same collection; selected paratypes in USNM.

Collection data.—California: from leaf mold, Dorrington, Calaveras Co.; leaf mold, Devil's Post Pile, Mono Co.; leaf mold, Howard Oak Bridge, Sierra Co.; pine leaf mold, Meyers, El Dorado Co.

Hemicheyletia rostella combines some of the features of the *bakeri* group of species with other features of *Paracheyletia pyriformis*. The characteristics of the gnathosoma of *H. rostella* and *P. pyriformis* are perhaps indistinguishable, but in the organization of tarsus I, the body setae, and the arrangement of dorsal plates (but not their surface ornamentation) *rostella* resembles *H. bakeri*. We are able to separate specimens of *rostella* from other *Hemicheyletia* species by its longer mouthparts and characteristically fashioned dorsal setae. *Paracheyletia pyriformis* is distinguishable from *Hemicheyletia rostella* by the form of its aberrant setae, the excessively lengthened axial paraterminal seta on tarsus I, and the six setae on tibia I. The retention of *pyriformis* in *Paracheyletia* creates a problem with regard to the generic assignment of *rostella*.

Three additional species are not redescribed or illustrated here. Two of them were included in De Leon's (1967) final paper.

Hemicheyletia anarbora (De Leon). New combination.
  Paracheyletia anarbora De Leon, 1967, Some Mites of the Caribbean Area, p. 34.

De Leon states that this species has a single pair of very small dorsomedian setae. His illustration (fig. 50, p. 31) shows one pair on the propodosoma and one pair on the hysterosoma. We have taken the figure to represent the true condition.

Hemicheyletia wellsina (De Leon). New combination.
  Paracheyletia wellsina De Leon, 1967, Some Mites of the Caribbean Area, p. 34.
Hemicheyletia bregetovae (Volgin). New combination.
  Dendrocheyla bregetovae Volgin, 1969, Akad. Nauk S.S.S.R., Zool. Inst., Opredel. p. Faune S.S.S.R. No. 101:207–208.

Paracheyletia Volgin

Paracheyletia Volgin, 1955, Akad. Nauk S.S.S.R., Zool. Inst., Opredel. p. Faune S.S.S.R. No. 59:168–169.
  Palp claw with 12 or 13 teeth. Palp tarsus with 2 combs and 2 sickles. Axial mouthparts noticeably elongate. Protegmen elevated or domed. Rostrum produced to a fairly acute point; length of its exposed part equals at least midline length of protegmen. Eyes prominent. Two plates entirely cover dorsum of idiosoma. Dorsal body setae of two kinds: dorsolateral setae fan-shaped, dorsomedian setae fragmented (in females). Tarsus I bears exceptionally long solenidion *w*I; guard seta absent or inconspicuous. Paired claws and empodia on tarsus I diminutive, borne on a short but well-defined pedicel; paraterminal setae solenidiform, both emphatically but unequally lengthened. Posteriormost anal setae acicular, smooth or nearly so.
  Type species—Cheyletus pyriformis Banks; by original designation.

Paracheyletia Volgin may be traced to a mistaken identification made by Oudemans (1906). Baker (1949) realized that Oudemans had confused two species in his effort to redescribe both sexes of *Cheyletia flabellifera* (Michael). Baker recognized that the male illustrated by Oudemans was in fact the male of *Cheyletia pyriformis* (Banks), 1904. At a later date, Volgin (1955) independently recognized Oudemans' mistake and used the male as the type of a new genus and species which he named *Paracheyletia assimilis*. However, since the type of Volgin's new genus, *Paracheyletia*, is the male of *Cheyletus pyriformis* (Banks), as noted by Baker, the name *assimilis* is a synonym of *pyriformis*. The type of this genus is, therefore, *Cheyletus pyriformis* Banks, 1904.

Baker (1949) transferred *flabellifer* of Michael, based on the female, to a new genus, *Eucheyletia*.

The writers believe that this genus has been used by other workers to receive species which differ significantly from the type. In this paper these species have been transferred to the genus *Hemicheyletia* Volgin. There are four characters which serve to distinguish *P. pyriformis* (Banks) and other species of *Parachey-*

letia from species of *Hemicheyletia* included herein, with the exception of *H. rostella,* which is intermediate in form, and will be discussed separately. The four characteristics are: (1) In *Paracheyletia* the palp claw has a larger number of teeth than are found in most *Hemicheyletia.* Although *H. congensis* may have up to 11 teeth, the more typical number in this genus is between 6 and 8. *Parachey-letia pyriformis* has 12 or 13 teeth on the palp claw. (2) The gnathosoma is notice-ably elongated in *Paracheyletia,* and the rostrum tends to be more pointed than in *Hemicheyletia.* The length of the exposed part of the rostrum in *Paracheyletia* exceeds the length of the protegmen. In *Hemicheyletia* it tends to be shorter than the length of the protegmen. Similarly, the length of the exposed part of the rostrum is about equal to the width of the protegmen in *Paracheyletia,* whereas it is much less so in *Hemicheyletia.* (3) In species of *Hemicheyletia* in which the dorsomedian setae are modified, the form of these setae is quite different from the form in *Paracheyletia.* In *Paracheyletia* they resemble the modified setae found in the genera *Mexecheles* and *Prosocheyla.* (4) The nature of the claws and the paraterminal setae of tarsus I in *P. pyriformis* are different from those of *Hemi-cheyletia.* In *pyriformis* there is a substantial pedicel, but the paired claws and empodium are very small. High magnification is required to confirm their presence. In addition, the paraterminal setae are solenidiform and are elongated (fig. 13*a*). The lateral paraterminal seta is at least twice as long as the pedicellar part of the tarsus, whereas the mesal paraterminal seta is only about half as long as the addorsal seta on the corresponding side.

Paracheyletia pyriformis (Banks)

(Fig. 13)

Cheyletus pyriformis Banks, 1904, Proc. U. S. Nat. Mus. 28:17; Banks, 1906, Proc. Ent. Soc. Wash. 7:135.

Cheyletia flabellifera (male) of Oudemans, 1906; Baker, 1949, Proc. U. S. Nat. Mus. 99(3228): 298–299. Male.

Cheyletus longipalpus Ewing, 1909; Baker, *ibid.*:298.

Paracheyletia assimilis Volgin, 1955, Akad. Nauk S.S.S.R., Zool. Inst., Opredel. p. Faune S.S.S.R. No. 59:168–169. New synonym.

Palp claw bears 12 or 13 teeth; dental ridge about two-thirds as long as claw. Palp tibia extended beyond tarsal joint to form a slender pedestal supporting claw. Outer comb ca. 20 teeth; inner comb ca. 25 teeth. Dorsal seta on palp femur and genu robust, lanceolate, with rows of barbs closely appressed. Protegmen domed, rounded and slightly incised in front, its cuticula roughened with a dense coating of coarse protruding tubercles; tubercles larger than those on tegmen. Tubercles covering tegmen aligned in ill-defined rows, interspersed with fine varicose striae; striations longitudinal on front portion, becoming horizontal over basal area. Rostrum slender, tapered to blunt point, bulged at level of pharyngeal pump; exposed part longer than protegmen; relative lengths of tegmen, protegmen, and rostrum = 16:7:9, respectively. Peritremes horseshoe-shaped, deeply inlaid, without acute flexures. Body plates cover entire dorsum; propodosomal and hysterosomal plates nearly equal in width, contiguous or overlapping at humeral sulcus; humeral plates of substantial size, longest diameter about equal to length of humeral setae, displaced to pleural or pleuroventral position; plates abundantly stippled with raised tubercles, these slightly larger near plate margins, somewhat smaller near midline and on hysterosoma; tubercles between median propodosomal setae tend to group into transverse rows. Eyes large, corneas clear, protuberant, encircled by 1 or 2 rows of broken striae. Dorsal setae: as many as 23 pairs plus humerals; 10 pairs broadly spatulate, hyaline blade portion flaring abruptly from short axial stalk, 5 to 7 coarsely barbed, sinuous ribs on convex face and about as many smooth, radially disposed ridges on concave undersurface; 9 to 13 pairs aberrant setae distributed as 5 or 6 pairs on propodosomal plate and 4 to 7 pairs on hysterosomal plate, these intermingled with conventional spatulate setae to obscure type and row relationships. Each aberrant seta consists of a short stalk and radial branches which thin out to become tenuous filaments of undefined matrix; each filament binds together a small number of crescentic sclerites (fig. 13c).

Length ratio: leg I/idiosoma = 0.7. Setae on legs I to IV: femora 2-2-2-1, genua 3-2-2-2, tibiae 6-5-4-4, tarsi 8-8-7-7. Tarsus I: pedicel very short, claws diminutive (ca. 4), about one-fourth as long as claws on tarsus II; paraterminal setae solenidiform, emphatically lengthened; mesal paraterminal about half as long as addorsals, lateral paraterminal nearly twice as long as pedicel and half as long as its mesal counterpart; solenidion wI very long (47); guard seta absent; ventral seta v smooth. Measurements (n = 1): length idiosoma 327, gnathosoma 125, leg I 236, tarsus I 74.

Collection data.—From lilac and poplar leaves, Franklin, Idaho (G. F. Knowlton and Shi-Chun Ma).

Volgin recently added three more species to *Paracheyletia:*

Paracheyletia hortensis Volgin, 1969, Akad. Nauk S.S.S.R., Zool. Inst., Opredel. p. Faune S.S.S.R. No. 101:181–182. Male.
Paracheyletia recki Volgin, 1966a, Akad. Nauk S.S.S.R., Trudy Zool. Inst. 37:283–285. Female.
Paracheyletia samsinaki Volgin, 1966a, *ibid.*:281–283.

Lepidocheyla Volgin

Lepidocheyla Volgin, 1963b, Entomol. Oboz. 42(4); transl. in Entomol. Rev. 42(4):506.

The description of the type species, *L. gracilis* Volgin, contains two characters which possibly are of generic significance, and tentatively serve to distinguish *Lepidocheyla* from *Hemicheyletia*. In this species there is a hysterosomal constriction between the metapodosoma and opisthosoma. The constriction is accompanied by corresponding marginal indentations in the hysterosomal plate. The hysterosoma is strongly tapered posterior to these marginal constrictions.

According to Volgin, the type species has nine small basal teeth on the palp claw, triangular scalelike setae, conventional dorsomedian setae (4 propodosomal and 3 hysterosomal), eyes, slender claws of equal size on all tarsi, peritremes with only four segments on each side, and a granulated pattern on the tegmen, protegmen, and idiosomal plates.

It is not possible on the basis of the writers' present knowledge of this genus to evaluate the generic significance of the characters given by Volgin. *Lepidocheyla* is treated here as a distinct genus closely allied to *Hemicheyletia*.

Hoffmannita Pelaez

Hoffmannita Pelaez, 1962, Anal. Escuela Nac. Ciencias Biol. II:72–73.
Myrmicocheyla Volgin, 1963b, Entomol. Oboz. 42(4); transl. in Entomol. Rev. 42(4):504. New synonym.

Palp claw with a single enlarged basal lobe. Palp tarsus bears 2 combs and 2 sickles. Protegmenal part of stylophore elevated, bulbous, covering a good part of rostrum. Eyes present or absent. Two unpaired plates cover entire dorsum of idiosoma. Dorsolateral body setae fan-shaped, ribbed and barbed. Dorsomedian setae somewhat modified, described as "navicular" (boat-shaped) or "petaloid," slender, delicate, numerous (9 to 15 pairs). Solenidion wI 1.8 to 2.0 times as long as greatest width of tarsus I. Guard seta barbed, longer than wI. Tarsi I to IV with paired claws and rayed empodia.

Type species.—*Hoffmannita mexicana* Pelaez; by original designation.

There are two very well described species:

Hoffmannita mexicana Pelaez, 1962, Anal. Escuela Nac. Ciencias Biol. II:73–77. Only known
  examples from scorpions, *Centruroides flavopictus,* in northeastern part of Oaxaca, Mexico.

Hoffmannita clavipes (Volgin), 1963*b*, Entomol. Oboz. 42(4); transl. in Entomol. Rev. 42(4):
  504–506. New combination. Taken from nest of *Formica rufa* L. in Russia, and on *Aradus* spp.,
  Russia and Czechoslovakia.

Although the dorsomedian setae of both species are similar in appearance, the
numbers of these are different. In *mexicana* there are only three pairs of these
modified setae on the hysterosomal plate. *H. clavipes* has eight pairs of modified
dorsomedians. *H. clavipes* is described as lacking eyes. Otherwise, it is very similar
to *H. mexicana.*

Laeliocheyletia, new genus

Palp claw with 4 or 5 teeth. Protegmen inconspicuous, depressed, merged with rostrum. Eyes
present. Two strongly sclerotized plates entirely cover dorsum and extend onto sidewalls; plates
overlapped at humeral sulcus; humeral platelets displaced to pleuroventral position. Dorsal body
setae stout, slightly curved rods, with inconspicuous flutes but heavy barbs. Dorsomedian setae
on both principal plates; dorsomedians and dorsolaterals similar in form. Anogenital apparatus
with a large postanal swelling which is wider than longitudinal length of genital covers. Legs
relatively short; leg I about 0.6 times as long as body. Solenidion wI slightly longer than greatest
diameter of tarsus; guard seta shorter than wI, smooth, in duplex position.

Type species.—Laeliocheyletia teretis, n. sp.; by present designation.

Laeliocheyletia teretis, new species

(Fig. 14)

Palp claw with 4 or 5 bluntly rounded teeth. Outer comb ca. 18 teeth, terminal tooth a broad
scimitar-like blade; inner comb ca. 20 fine teeth. Rostrum broad, apex rounded, length of exposed
part in front of peritremes almost equal to length of tegmen. Protegmen depressed, without surface ornamentation; margins converge to fit into rostral gutter. Tegmen decorated with fretwork
of varicose anastomosing ridges or striae and about 30 small, widely dispersed pores. Peritremes
circumscribe sculptured area of tegmen, 6 well-defined links per side. Two subequal plates cover
entire dorsum of idiosoma, overlapped at humeral sulcus, each transparent, weakly sclerotized,
with overlay of very fine bacillus-like broken striae. Humeral plates displaced to ventrolateral
position, wedged between coxae II and III. One pair of eyes. Dorsal setae stout fluted rods,
moderately long, each with 4 or more rows of short, stout barbs tightly appressed to shaft along
distal half of its convex surface; 16 pairs excluding humerals, 4 and 6 dorsolaterals, 3 and 3
dorsomedians; all similar in form; length of various dorsomedians nearly uniform (55–66);
length of dorsomedians exceeds dorsolaterals (31–51), noticeably more so on hysterosoma. Setae
on coxae and venter of body acicular, smooth, longest less than twice as long as shortest. Anal
orifice appears to be a transverse slit partly shielded by outcurved posterior lips of genital covers;
a large postanal swelling bears 3 pairs of simple anal setae; posterior rim of this bulbous protuberance comprises a sclerotized shell supported by a forked apodome, arms of fork short,
pointed, divergent, anteriorly directed.

Legs stubby. Length ratio: leg I/*idiosoma* = 0.6, legs I/II = 1.2. Setae on legs I to IV: femora
2-2-2-1, genua 3-2-2-2, tibia 5-4-4-4, tarsi 10-8-7-7. Tarsus I: solenidion wI slender, moderately
long (20); guard seta (8) smooth, less than half as long as wI; solenidion wI and guard seta
appear to share same socket. Measurements (holotype): length idiosoma 304, gnathosoma 109,
leg I 193, leg II 148, tarsus I 86.

*Holotype.—*Female, from prosternum of tenebrionid beetle taken from debris in an ant nest,
Quezaltepeque, El Salvador (D. Cavagnaro, M. Irwin); deposited in UCD.

Ker Muma

Ker Muma, 1964, Florida Ent. 47(4):250.

Small mites with relatively short legs. Palp claw edentate. Palp tarsus with 2 sickles and 2 combs. Eyes present. Two principal body plates and stylophore coarsely reticulate. Dorsal body setae spatulate or fanlike; dorsomedians and dorsolaterals alike in form. Genital and anal covers lie close together to form a broad, ovoid assembly; 2 pairs of paragenital setae, 2 pairs of genitals, 3 pairs of anals; 1st and 2d anals set in a straight crossrow; 3d anals plumose or fanlike, set behind 1st and 2d pairs. Solenidion wI about one-third as long as tarsus; its nipple arises considerably anterior to base of pinnate guard seta. Smooth claws and multirayed empodia on legs I–IV.

Type species.—Ker palmatus Muma, 1964; by original designation.

Ker palmatus Muma
(Fig. 15)

Ker palmatus Muma, 1964, Florida Ent. 47(4):250–252.

Outer comb ca. 15 teeth; inner comb ca. 20 teeth. Dorsal seta of palp tibia set on a shallow tubercle, not on a mesal lamella present in many other cheyletids. Dorsal seta of palp femur spatulate, tapered to a rough point; upper seta on genu palmate. Protegmen elevated, concave on front margin; accounts for about one-third of midline length of stylophore. Tegmen and protegmen with a few meshes of a coarse reticulum; each cell of meshwork encloses numerous flocculent varicosities or faint, broken striae; this pattern repeated on main body plates.

Hysterosomal plate with 2 pairs of marginal notches between dorsolateral setae 1-2 and 2-3. Dorsal body setae: 15 pairs plus humerals; laterals and medians structurally similar (16–24); shape of each seta fan-shaped to flabellate, somewhat like a squash racquet, with shallow ribs projecting beyond margin of blade proper to produce a fringe of radially disposed spikelets. Propodosomal plate bears 4 pairs of marginals plus 3 pairs of dorsomedians. Posterior pair of anal setae flabellate.

Length ratio: leg I/idiosoma = 0.6. Setae on legs I to IV: femora 2-2-2-1, genus 3-2-2-2, tibiae 6-4-4-4, tarsi 9-8-7-7. Tarsus I with very short pedicel; infraterminal setae on body of tarsus rather than on pedicel; solenidion wI (16) nearly half as long as pinnate guard seta (31); nipple of wI distal to alveolus of guard seta. Measurements (holotype): length idiosoma 228, gnathosoma 70, leg I 133, tarsus I 51.

*Collection data.—*Known only from type locality: leaf litter, Sebring, Florida.

Ker bakeri Zaher and Soliman
(Fig. 16)

Ker bakeri Zaher and Soliman, 1967, Bull. Soc. Ent. Egypte 51:24–25.

Outer comb ca. 15 teeth; inner comb ca. 20 teeth. Dorsal seta on femur and genu of palp narrow spatulate. Stylophore bullet-shaped; protegmen subconic, elevated, with a shallow notch at apex, no obvious surface ornamentation; tegmen faintly marked with a coarse reticulum. Peritremes with 3 or 4 slender links per side. Eyes present but inconspicuous. Two dorsal plates cover most of idiosoma, both reticulate. Hysterosomal plate with a pair of notches between dorsolateral setae 1-2, and dorsolaterals 2-3; all dorsal hysterosomal setae appear to be implanted on this plate. Dorsal body setae: 15 pairs plus humerals (23–31; humerals 39); lateral and median setae similar in form; blade of each seta spatulate with a rounded end and fringe of fine barbs and very delicate ribs. With 4 pairs of dorsolateral setae plus 2 pairs of dorsomedian setae on propodosomal plate; 6 pairs of dorsolaterals plus 3 pairs of dorsomedians on hysterosomal plate (penultimate pair of lateral hysterosomals—actually aligned with medians in this instance—are classed as 5th dorsolaterals).

Length ratio: leg I/*idiosoma* = 0.5. Setae on legs I to IV: femora 2-2-2-1, genua 3-2-2-2, tibiae 5-4-4-4, tarsi 9-8-7-7. Anogenital region as illustrated (fig. 16*b*). Tarsus I: pedicel bears only 2 paraterminal setae; 2 infraterminals set back on main part of tarsus; mesal addorsal seta (47) more robust and longer than its lateral mate (27); nipple of *w*I arises about in middle of tarsus; guard seta (35) lanceolate, arising considerably behind *w*I. Measurements (n = 1): length idiosoma 281, gnathosoma 74, leg I 148, tarsus I 62.

Collection data.—Illustrated specimen taken from a bird nest, Giza, Egypt.

This species is larger than *palmatus* and its dorsal plating has no obvious porosities or microtubercles. The dorsal body setae are narrow spatulate rather than fan-shaped. Tibia I has six setae instead of five as in *palmatus*. The distal solenidion on tibia I is very minute and does not reach to the end of its segment; in *palmatus* this solenidion is inflated and long enough to project beyond the tibiotarsal flexure by at least half its length.

The specimen examined in the USNM collection is possibly a paratype, although not so marked.

Pavlovskicheyla Volgin

Pavlovskicheyla Volgin, 1965, Akad. Nauk S.S.S.R., Trudy Zool. Inst. 35:290–291.

Palp claw edentate; palp tarsus with 2 combs and 2 sickles. Protegmen elevated, extending considerably in front of peritremes; anterior rim excavate or slightly concave. Eyes present. Two dorsal shields cover idiosoma. Dorsal idiosomal setae lanceolate, ribbed and barbed; laterals and medians similar. With 4 pairs of laterals plus 2 pairs of medians on propodosomal shield, 6 plus 3 pairs on hysterosoma. First laterals originate on platelets imperfectly separated from principal plate by 4 rows of striae. Solenidion *w*I on tarsus I is half as long as this segment; guard seta longer than *w*I, acicular, finely barbed. Smooth claws on all tarsi.

Type species.—*Cheletophyes semenovi* Rohdendorf, 1940, Moscow Univ. Uchenyi zapiski, Wissensch. Ber. 42: 94, 96; by original designation.

Volgin (1965) redescribed the type species as *Pavlovskicheyla semenovi* (Kuzin). Rohdendorf (1940) published the first description, but he too attributed the authorship to Kuzin because he wished to acknowledge the use of materials and unpublished notes supplied by the latter.

The combination of characters ascribed to *Pavlovskicheyla* does not sharply differentiate this genus from *Hemicheyletia* or *Ker*. As more new species are discribed in each of these three genera, it will be possible to assess more critically the worth of the generic characters involved in the present arrangement.

Cheyletus Latreille

Cheyletus Latreille, 1796, Précis des caractères génériques des insectes. . . . An. 179.

Eutarsus Hessling, 1852, Illustr. Med. Zeit. 1(5):258; Baker, 1949, Proc. U.S. Nat. Mus. 99(3238):275–276.

Cheletes Latreille (emend. nomen Oudms.), Oudemans, 1903c, Tijdschr. Ent. 46:121; 1906, Mém. Soc. Zool. France 19:78–79.

Palp tarsus carries 2 prominent sickles and 2 comblike setae having numerous tines. One to several blunt teeth on base of palp claw. Protegmen subconic, merged, sufficiently developed to accommodate basal sclerites of cheliceral stylets, but otherwise its structure not easy to distinguish from underlying rostrum. Tegmen plain or lightly striated, with prominent marks indicating areas of muscle attachment. Peritremes often M-shaped and acutely bent where transverse arms turn rearward. Propodosomal plate covers most of propodosoma. Eyes absent. Hysterosomal plate tends to be quadrangular in shape, usually broadest on front margin, about two-thirds as wide as propodosomal plate, incompletely covering tergal area of hysterosoma. Dorsolateral body setae acicular, fusiform or very narrow spatulate, comparatively short; usually 10 pairs plus humerals: 4 pairs on propodosomal plate, 2 to 5 pairs on hysterosomal plate. Dorsomedian setae present or absent, usually very small and inconspicuous, when present usually aberrant in form, most species with 1 to 3 pairs. Tibiae I to IV have 1 dorsal seta much longer than other setae on this segment. Addorsal setae *tc* on tarsus I roughened or minutely annulate. All tarsi with paired claws and rayed empodia.

Type species.—Acarus eruditus Schrank, 1781; by monotypy.

The key below covers only the females of nine species for which study material was available in the United States. The names of twenty-seven other species and the journals in which the original descriptions appeared are listed at the end of this section. Owing to the loss of type material and inadequate descriptions by earlier workers, some of these twenty-seven species may never be dealt with constructively. Others in the list may prove to be junior or senior synonyms of one or more of those which we have here identified. A complete revision of this classical genus may require the combined efforts of several observers having access to types from all parts of the world.

KEY TO FEMALES OF CHEYLETUS

1. Guard seta of tarsus I absent or obviously shorter than solenidion *w*I 2
   Guard seta of tarsus I present and obviously longer than solenidion *w*I 5
2. Palp claw with 1 large basal tooth ....*fortis* Oudemans
   Palp claw with 2 or more basal teeth 3
3. Femur IV with 2 setae ....*eruditus* (Schrank)
   Femur IV with 1 seta 4
4. Midline distance separating propodosomal and hysterosomal plates exceeds length of 1st
      dorsolateral seta on hysterosoma ....*malaccensis* Oudemans
   Midline distance separating these plates much less than length of 1st dorsolateral seta on
      hysterosoma ....*malayensis* Cunliffe
5. With 1st dorsolateral seta on hysterosoma arising on principal hysterosomal plate
                                                        cacahuamilpensis Baker
   With 1st dorsolateral seta on hysterosoma arising on individual, trivial platelet, 1 platelet
      for each seta of this pair 6
6. With 8 pairs of small vesicular or squamous median setae ....*linsdalei* Baker
   With not more than 3 pairs of small vesicular or squamous median setae 7
7. Palp claw with 2 basal teeth ....*aversor* Rohdendorf
   Palp claw with 3 or 4 basal teeth 8
8. Dorsolateral body setae flattened to form a thin, narrow, fusiform or parallel-sided blade
      bilaterally fringed with slender barbs; no dorsolaterals flagelliform ...*trouessarti* Oudemans
   Dorsolateral body setae not noticeably flattened; they appear to have a solid core sheathed
      with barbs; some dorsolaterals flagelliform ....*hendersoni* Baker

Cheyletus eruditus (Schrank)

(Fig. 17)

Acarus eruditus Schrank, 1781, Enumeratio insectorum Austriae indigenorum, p. 513.
Eutarsus cancriformis Hessling, 1852; Oudemans, 1938.
Cheyletus seminivorus Packard, 1878; Beer and Dailey, 1956.
Cheyletus eburneus Hardy, 1867; Baker, 1949.
Cheletes eruditus (Schrank), Oudemans, 1903c.
Cheyletus ferox Banks, 1906; Baker, 1949.
Cheyletus rabiosus Rohdendorf, 1940; A. M. Hughes, 1961.
Cheyletus butleri A. M. Hughes, 1948, 1961.
Cheyletus doddi Baker, 1949. New synonymy.
Cheyletus eruditus (Schrank); Oudemans, 1903c, 1906; Newstead and Duval, 1918; Rohdendorf,
   1940; Haarlov, 1942; A. M. Hughes, 1948, 1961; Baker, 1949; Beer and Dailey, 1956.

Palp claw with 2 similar basal teeth. Dorsal palptibial seta acicular, smooth. Outer comb ca. 13 teeth; inner comb ca. 15 teeth. Dorsal seta of palp femur long (ca. 130), sparsely barbed on distal half. Tegmen ornamented with broken striae and anastomosing trabeculae between origins of muscle bundles; striae spread fanwise from base of tegmen; a few heavy striae or cuticular ridges interspersed among more numerous rows of delicate, interrupted striae. Protegmen indistinct, not obviously ornamented. Peritremes M-shaped, 4 to 5 links in descending arms; transverse arms curve forward to acute anterolateral flexures. Dorsal plating plain, no surface sculpturing other than outlines of attached muscle bundles and very minute scattered pores. Propodosmal plate nearly coextensive with body outline. Hysterosomal plate approximately trapezoidal, front margin arched, incompletely covering opisthosoma. A wide belt of transversely striated integument separates propodosmal and hysterosomal plates. Dorsolateral setae acicular, finely barbed; 10 pairs, alike in form, unequal in length (51–70); 4 pairs on outer curved margins on propodosomal plate; 1st pair of dorsolaterals of hysterosoma on minute individual platelets (fig. 17a); next 3 pairs of margins of hysterosomal plate; 2 pairs subterminal, each seta on its own platelet. No dorsomedian setae. Humeral seta acicular, sparsely barbed, relatively long (117–164).

Length ratio: leg I/idiosoma = 0.9. Distribution of setae on legs I to IV: femora 2-2-2-2, genua 3-2-2-2, tibiae 6-4-4-4, tarsi 10-8-7-7. Seta *dt* on tibiae III (121) and IV (147) relatively long, smooth, flagelliform. Tarsus I: solenidion *w*I short (27); guard seta tiny half as long as *w;* addorsal setae roughened or minutely annulate. Measurements (n = 10): length idiosoma 465 ± 53, gnathosoma 186 ± 13, leg I (movable segments) 410 ± 36, tarus I 147 ± 8.

Collection data.—California: floor of dairy barn, Davis; in eucalyptus bark, Burlingame; from moss, Spanish Flat, Napa Co.; haybarn litter, St. Helena, Napa Co., and Hallwood, Yuba Co.

Cheyletus malaccensis Oudemans

(Fig. 18)

Cheletes malaccensis Oudemans, 1903b, Ent. Ber. Nederl. Ver. 1 (12) :84; Oudemans, 1906, Mém. Soc. Zool. France 19:88–96.

Cheyletus malaccensis, Baker, 1949, Proc. U. S. Nat. Mus. 99(3238):284–285; A. M. Hughes, 1961, Ministry Agric., Fish. and Food Tech. Bull. 9:197–200.

Cheyletus munroi Hughes, A. M., 1948, 1961.

Palp claw bears 2 unlike basal teeth, distal tooth a rounded cone, longer than rectangular proximal tooth. Outer comb ca. 20 teeth; inner comb ca. 30 teeth. Dorsal seta of palp femur very long (160), flagelliform, with few very fine barbs. Palp femur robust, completely striated, extending slightly beyond anteriormost bend of peritremes. Tegmen with fine, closely set striae disposed fanwise from tegmenal base line, sharply outlined area of muscle attachment and numerous distinct pores. Striae and pores of tegmen continue onto protegmen but fade from view a short distance in front of peritremes. Peritremes M-shaped; transverse arms curve forward, then flex sharply backward at acute angle; descending arms with 6 or 7 links. Dorsal plating very delicately striated, the striae broken into very short, wavy sections; small pores most numerous near 1st 3 pairs propodosomal setae. Propodosomal plate covers most of propodosoma; hysterosomal plate approximately rectangular, covering midsection of opisthosoma. A wide belt of transverse striae separates propodosomal and hysterosomal plates. Dorsolateral setae narrow, flattened or rolled blades, fusiform in outline, some tapered to sharply pointed tips, completely fringed with slender barbs; 10 pairs (70–94), situated as in *eruditus*. Dorsomedian setae generally absent but may occur sporadically in asymmetrical arrangement, e.g., 2 on one side of propodosomal plate; when present they are very small flat blades, narrow, frayed at ends. Humeral setae acicular, densely barbed, finely pointed, relatively long (121–179).

Length ratio: leg I/idiosoma = 0.8. Distribution of setae on legs I to IV: femora 2-2-2-1, genua 3-2-2-2, tibiae 6-4-4-4, tarsi 10-8-7-7. Setae *dt* on tibiae III (142) and IV (189) very long, smooth. Tarsus I: solenidion *w*I short (22), tapered and pointed; guard seta very short and fine, difficult to observe; addorsal setae *tc* roughened or minutely annulate. Measurements (n = 9): length idiosoma 585 ± 75, leg I 441 ± 39, gnathosoma 201 ± 12, tarsus I 169 ± 15.

Collection data.—Feed trash, swine barns, U. C. campus, Davis; rodent cages, School of Veterinary Medicine, U. C., Davis; red rice, Indonesia.

Cheyletus aversor Rohdendorf

(Fig. 19)

Cheyletus aversor Rohdendorf, 1940, Moscow Univ. Uchenye zapiski, Wissensch. Ber. 42:86–87;
Volgin, 1969.
Cheyletus beauchampi Baker, 1949, Proc. U. S. Nat. Mus. 99(3238):282–283. New synonym.

Palp claw with 2 dissimilar basal teeth; distal tooth more conical, longer than proximal tooth.
Outer comb on palp tarsus ca. 17 teeth; inner comb ca. 30 teeth. Palp femur robust, extending
but slightly beyond basal sclerites of cheliceral stylets; dorsal seta long (95), barbed from base
to tip; barbs short, fine, numerous. Tegmen with a few faint linear ridges, numerous minute
pores. Protegmen not obviously ornamented. Peritremes with ca. 8 links per side, proximal segments transverse, anterolateral flexure obtuse. Dorsal plating plain, with numerous small scattered pores. Propodosomal plate covers most of propodosoma; hysterosomal plate almost rectangular, incompletely covering opisthosoma; these two plates separated by a broad belt of transverse striae. Dorsolateral setae flattened, narrow, spatulate, bilaterally fringed; 10 pairs, somewhat variable in length (51–74): 4 pairs on outer perimeter of propodosomal plate, 1 pair on
integument in midregion of body, 3 pairs on margins of hysterosomal plate, 2 pairs on caudal
integument. Dorsomedian setae: 2 pairs, very short, ill-defined paddles or vesicles: 1 pair on
propodosomal plate almost aligned with posterior marginals of that plate, 1 pair on integument
close to front margin of hysterosomal plate. Humeral setae acicular, abundantly barbed, long
(90).

Length ratio: leg I/idiosoma = 0.7. Distribution of setae on legs I to IV: femora 2-2-2-1, genua
3-2-2-2, tibiae 6-4-4-4, tarsi 10-8-7-7. Setae dt on tibiae III (70) and IV (99) noticeably longer
than others on these legs, appreciably barbed. Tarsus I: solenidion wI short (27); guard seta
(49) almost twice as long as wI, acicular, plain or with 2 or 3 basal barbs; addorsal setae tc
roughened or minutely annulate; ventral seta v conspicuously barbed, plumose. Measurements
(n = 10): idiosoma 492 ± 43, gnathosoma 163 ± 9, leg I 328 ± 24, tarus I 119 ± 7.

Collection data.—Only one population found, in floor litter, animal exhibition barn, U. C.
campus, Davis, Calif.

Although *aversor* resembles *eruditus* in many respects, there are numerous
differences of which these are most readily determined: the dorsolateral body setae
are flat, narrow spatulate, and fringed with barbs; there are two pairs of small,
modified dorsomedian setae; femur I has only one seta (two in *eruditus*); the
rearward flexure of each peritreme is not at an acute angle; the guard hair on
tarsus I is nearly twice as long as solenidion wI.

Specimens independently identified as *C. aversor* Rohdendorf, from Davis,
California, match perfectly a specimen from Czechoslovakia in the USNM collection which was identified as *C. aversor* Rod. by Dr. Karel Samšiňák.

Cheyletus cacahuamilpensis Baker

(Fig. 20)

Cheyletus cacahuamilpensis Baker, 1949, Proc. U.S. Nat. Mus. 99(3238):282.
Eucheyletia mungeri McGregor, 1956, Mem. So. Calif. Acad. Sci. 3(3):24. New synonymy. (Holotype: Los Angeles County Museum, No. 68)

Palptibial claw with 2 basal teeth, distal tooth more massive and longer than basal tooth.
Outer comb ca. 18 teeth, the terminal tooth (i.e., tip) extended as a long, curved, pointed spine
more than twice as long as tooth next behind. Inner comb ca. 32 teeth. Dorsal seta of palp femur
relatively short (55), slightly flattened, abundantly barbed. Palp femur short, robust, reaching
slightly beyond marginal flexure of peritreme, entirely striated; striae fragmented and whorled
in small area beside dorsal seta. Tegmen with indistinct granulose roughening, marked with
5 or 6 anastomosing ridgelike folds. Protegmen not obviously ornamented. Peritremes M-shaped,

sharply flexed, 7 or 8 links in descending arm. Inferior adoral setae on tip of rostrum longer than superior adorals, ratio of lengths 10 to 8, respectively. Dorsal plating with faintly pebbled or granular surface texture. Propodosomal plate completely covers propodosoma; hysterosomal plate quadrangular, about two-thirds as wide as propodosomal plate and separated from it by a narrow band of striae. Dorsolateral setae narrow spatulate, barbed, with rounded ends, fluted on convex face, slight curved, 1 to 3 barbed ribs; 10 pairs: 4 pairs on margins of each plate, 2 pairs terminals on integument. Dorsomedian setae appear as diminutive collapsed pyriform bladders; 2 pairs, 1 pair on each dorsal plate. Humeral setae resemble neighboring dorsolaterals; each on an oval platelet about 0.7 times as long as setae it bears.

Length ratio: leg I/idiosoma = 0.9. Distribution of setae on legs I to IV: femora 2-2-2-1, genua 3-2-2-2, tibiae 6-4-4-4, tarsi 10-8-7-7. Seta *dt* on tibiae III and IV spatulate, barbed, not noticeably longer than opposite dorsal seta. Tarsus I: solenidion *w*I very short (12); guard seta long (60), acicular, fringed; addorsal setae *tc* roughened or minutely annulate. Measurements: length idiosoma 381 ± 23, leg I 273 ± 28, gnathosoma 148 ± 11, tarsus I 108 ± 8.

Collection data.—California: soil litter under black walnut trees, Brea Canyon, Los Angeles Co.; topsoil, Capay, Yolo Co.; topsoil, Jacolitos Canyon, Fresno Co.; in bat guano, Wilbur Springs mine, Colusa Co.; soil with Bermuda grass and weeds, Nicholas, Sutter Co.; topsoil, Grizzly Island, Solano Co.; topsoil, Winters; forest leaf mold, Pepperwood, Humbolt Co.; clay soil, Robbins, Yolo Co.; soil under barley straw, Woodland; *Pyracantha* leaf litter, Davis; trunk cavity in willow tree, U. C. campus, Davis; greenhouse floor, U. C. campus, Davis; soil beneath poultry cages, Davis; nest of *Bombus* sp., Berkeley; oak mulch, Colfax.

This species can be distinguished from its close allies according to the following features or certain combinations of them: the dorsal seta of the palp femur is short and barbed, the dorsolateral and humeral setae are narrow, spatulate, with four pairs on the hysterosomal plate; the presence of two pairs of small dorsomedians is constant among numerous specimens examined; the outer comb of the palp tarsus ends in a long curved, pointed spine; there are no ultralong setae on the dorsal face of tibiae III and IV; solenidion *w*I is a minute peg and the companion guard seta is long and barbed; the inferior adoral setae on the rostrum are noticeably longer than the superior adoral setae.

Cheyletus cacahuamilpensis is more cosmopolitan than the type habitat suggests. It was originally described from bat guano in a cave, in Mexico. It now appears that the species occurs in very diverse habitats and is frequently encountered in California. Specimens have also been recovered from bat guano in California, and thus the species shares this situation with *Cheletonella vespertilionis*. So far it has not been recovered from grain trash or sweepings in feed lots for domestic mammals.

Cheyletus trouessarti Oudemans

(Fig. 21)

Cheyletus trouessarti Oudemans, 1903a, Tijdschr. Nederl. Dierk. Ver. (ser. 2) 8:16.
Cheletes trouessarti Oudemans, 1903c, Tijdschr. Ent. 46:129–132; Oudemans, 1906, Mém. Soc. Zool. France 19:88.
Cheyletus trouessarti Rohdendorf, 1940, Moscow Univ. Uchenye zapiski, Wissensch. Ber. 42:88; A. M. Hughes, 1961, Ministry Agric., Fish. and Food Tech. Bull. 9:194–197.
Cheyletus davisi Baker, 1949, Proc. U. S. Nat Mus. 99(3238):283–284. New synonymy.

Palp claw usually with 3, occasionally 4, basal teeth. Outer comb on palp tarsus ca. 15 teeth; inner comb ca. 20 teeth. Dorsal seta on palp femur acicular, long (100), flagelliform, slightly barbed except near finely attenuated tip. Palp femur robust, extending slightly beyond flexure of peritreme, surface entirely striate. Tegmen striated fanwise; origins of muscle bundles clearly circumscribed; small irregularly scattered pores in cuticula expect where muscles attach; no prominent ridges among striae. Protegmen without obvious surface ornamentation. Peritremes acutely bent, 4 or 5 links in descending arms. Dorsal plating plain except for usual subcuticular irregularities. Propodosomal plate covers most of propodosoma; hyterosomal plate about 0.6 time as wide as front plate, broadest between anteriormost setae. A wide band of striae separates propodosomal and hysterosomal plates. Dorsolateral setae flattened to form narrow spindle-shaped blades fringed with finely pointed barbs; usually with no ribs or fluting; 10 pairs plus humerals, some slightly longer than others (35–51), arranged as in *eruditus*. Dorsomedian setae very small collapsed vesicles, thin-walled, inconspicuous; 1 pair on propodosomal plate, 2 pairs on hysterosomal plate, front pair distant from rear pair. Humeral setae less than twice as long as nearby dorsolaterals, variable in length (31–82), with fine barbs.

Length ratio: leg I/idiosoma = 0.7. Distribution of setae on legs I to IV: femora 2-2-2-1, genua 3-2-2-2, tibiae 6-4-4-4, tarsi 10-8-7-7. Setae dt on tibiae III and IV acicular, flagelliform, sparsely barbed, 68 and 88, respectively, longer than other setae on these legs. Tarsus I: solenidion wI short (20); guard seta about twice as long as wI, acicular, smooth; addorsal setae tc roughened or minutely annulate. Measurements (n = 10): length idiosoma 381 ± 37, leg I 273 ± 28, gnathosoma 135 ± 11, tarsus I 105 ± 8.

Collection data.—California: in bark of ponderosa pine, Yosemite Valley; soil beneath willow tree, Davis; dairy barn, U. C. campus, Davis; beneath haystack, Davis; topsoil, Olema, Marin Co.; topsoil, Elmira, Solano Co.; nests of *Neotoma* sp., Oakville, Modjeska Canyon; eucalyptus bark, Burlingame; pine forest duff, Pine Grove, Amador Co.

It is possible that *trouessarti* and *aversor* may be confused. The differentiating characters of *trouessarti* are: 3 or 4 basal teeth on the palp claw, about 20 teeth on the inner comb of the palp tarsus, two pairs of dorsomedian setae on the hysterosomal plate; the peritremes have an acute angle of flexion and, typically, four links in their descending arms. *C. aversor* has 2 basal teeth (typically) on the palp claw, about 30 teeth on the inner comb, one pair of dorsomedian setae on the hysterosomal plate; the peritremes have an obtuse angle of flexion and, typically, six links in their descending arms.

The five species of *Cheyletus* occurring locally near Davis, California, i.e., *eruditus*, *malaccensis*, *aversor*, *cacahuamilpensis*, and *trouessarti*, have been found in very similar habitats. Little or nothing is known about their particular ecological requirements or food preferences. Single samples of barn litter often contain more than one species. The anatomical variations by which the above-named species are separated may present some difficulty when additional and allied species are encountered.

Cheyletus fortis Oudemans
(Fig. 22)

Cheyletus fortis Oudemans, 1904*b*, Ent. Ber. Nederl. Ent. Ver. 1(18):161; Oudemans, 1906, Mém. Soc Zool. France 19:96–99; Baker, 1949, Proc. U.S. Nat. Mus. 99(3238):280–281.

The occurrence of a single, rounded toothlike process on the palp claw is the feature by which *fortis* can be identified. Other than this, we have not been able to find any meaningful quantitative or structural differences between this species and *malaccensis* Oudemans. The most variable of the body setae seems to be the humeral seta; typically in *fortis* it is flagelliform, long (to 175), and sparsely barbed. In samples from different localities this seta can become shorter, thicker, and more densely barbed until it resembles the nearby dorsolaterals.

Collection data.—Specimens examined in USNM collection included the following, most of them intercepted in quarantine inspections: on the skin of parakeet, New Guinea; on *Rattus coxinga* and *Callosciurus erythraeus,* Ma-an, Formosa; on *Callosciurus* sp., Burma; on tulip bulbs, garden beans, rice straw mat, Japan; on *Delphinium ajacis,* Australia; on *Allium sativum,* Philippine Islands.

Cheyletus linsdalei Baker

(Figs. 23*b, c*)

Cheyletus linsdalei Baker, 1949, Proc. U.S. Nat. Mus. 99 (3238):281.

Palp claw with 3 or 4 teeth. Outer comb ca. 18 teeth; inner comb ca. 24 teeth. Dorsal seta on palp femur and genu long, flagelliform, barbed. Peritremes M-shaped, 8 or 9 links per side. Dorsal plating without obvious surface ornamentation. Dorsal body setae: 18 pairs (probably) plus humerals; dorsolaterals and dorsomedians dissimilar. Dorsolateral setae narrow spatulate, with margins bearing numerous slender barbs; blades may be as wide as 20 percent of length (35–47), with 2 upper ribs and a greater number of thin-lined lower ribs; dorsolaterals 2-3-4 borne on hysterosomal plate, other dorsolaterals on individual platelets. Humeral setae also narrow spatulate, somewhat longer (59–66) than dorsolaterals. Dorsomedian setae appear to be small, exceedingly thin squames or collapsed bladders; 3 pairs on propodosomal plate, possibly 5 pairs on hysterosomal plate. Measurements (holotype only): length idiosoma 365, gnathosoma 125, tarsus ɪ 100, solenidion *w*I 27, guard seta 35.

Collection data.—Three type specimens collected on *Citellus beecheyi,* Monterey Co., Calif. Deterioration of these PVA-mounted specimens has obscured much of their finer organization.

The unique character of *linsdalei* is the occurrence of eight pairs of small, squamous dorsomedian setae. Otherwise, the 3 or 4 teeth on the palp claw, the relatively broad blades of the spatulate dorsolateral and humeral setae, and the lengths and spatial relations of *w*I and its guard seta are helpful in distinguishing this species.

Cheyletus hendersoni Baker

(Figs. 23a, d)

Cheyletus hendersoni Baker, 1949, Proc. U. S. Nat. Mus. 99(3238):279.

As far as can be determined from available specimens, *hendersoni* Baker is a "long-haired" relative of *trouessarti* Oudemans. The type slide, USNM 1757, contains several specimens, none of which are fully transparent or relaxed. It cannot now be affirmed with certainty that subsequently identified specimens are perfectly matched to the type. This species may be very troublesome to identify until it becomes possible to fix objective limits for intraspecific variations.

The specimens identified as *hendersoni* resemble *trouessarti* in most respects, including three pairs of tiny vesicular dorsomedian setae, with one pair on the propodosomal plate and two pairs on the hysterosomal plate. But in *trouessarti* the dorsolateral body setae are flattened, spindle-shaped blades fringed with finely pointed barbs. The corresponding setae of *hendersoni* appear to have a solid needle-like core entirely sheathed with barbs. Some of the dorsolateral setae are quite long and flagelliform.

Collection data.—Specimens taken from nests of *Neotoma albigula,* Santa Fe, New Mexico; debris in nest of *Colaptes cafer,* Cache la Poudre River, 2.5 miles E. of Fort Collins, Colorado.

Cheyletus malayensis Cunliffe

(Fig. 23*e*)

Cheyletus malayensis Cunliffe, 1962, Proc. Ent. Soc. Wash. 64(3):201–202.

Gnathosomal parts very closely resemble those of *malaccensis.* Outer comb ca. 16 teeth; inner comb ca. 20 teeth. Hysterosomal plate unusually large, covering most of tergal area; a narrow belt of striated integument separates hysterosomal and propodosomal plates. Dorsal body setae: 10 pairs plus humerals; no median setae observed. Each dorsolateral seta is a narrow, flattened blade with sides parallel, fringed with finely pointed barbs. Certain body setae noticeably long (range for 2 specimens): 1st propodosomals 86–90, 3d propodosomals 113–140; 2d hysterosomals 98–105. With 1st hysterosomal seta on its own platelet close beside anterolateral corner of hysterosomal plate; hysterosomals 2–4 inclusive on margins of this plate. Solenidion wI (27) and minute guard seta approximately in midsection of relatively short tarsus. Measurements (n = 1): length idiosoma 446, tarsus I 117.

Collection data.—C. malayensis known only from nests of birds in Malaya: *Munis atricapilla, M. striata,* and *Pycnonotus goiaver.*

The female of this species has no single distinguishing feature. The character which may be most useful for separating *malayensis* from kindred species having no dorsomedian setae is the width and extent of the hysterosomal plate. This plate extends forward to the level of the first hysterosomal setae, although the latter are set on their own trivial platelets. Otherwise, the lengths of certain dorsolateral setae give the impression that the species has unusually long setae. The observed ranges of lengths for three different setae, as given above, lie outside the observed ranges of small samples of corresponding setae on *eruditus* and *malaccensis*. The lengths of the humerals and first hysterosomals do not appear to separate these three species.

Cheyletus is a large genus and many species are recorded in literature. Those not available for restudy at this time are listed below in alphabetical order.

Cheyletus acarophagus Zaher and Soliman, Bull. Soc. ent. Egypte 51:25–26.

Cheyletus acer Oudemans, 1904*b*, Ent. Ber. Nederl. Ent. Ver. 1(18):162. Male.

Cheyletus alacer Oudemans, 1904*b*, *ibid.*:162. Male.

Cheyletus audax Oudemans, 1904*b*, *ibid.*:162. Male.

Cheyletus baloghi Volgin, 1969, Akad. Nauk S.S.S.R., Zool. Inst., Opredel. p. Faune S.S.S.R. No. 101:116.

Cheyletus burmiticus[3] Cockerell, 1917, Psyche 24:41.

Cheyletus carnifex Zachvatkin, 1935, A Short Key to the Granary Mites, p. 27.

Cheyletus clavispinus[3] Banks, 1902, Canadian Ent. 34:172.

Cheyletus ferox Trouessart, 1885, Bull. Soc. Etude Sci. d'Angers 14:90–91.

Cheyletus furibundus Rohdendorf, 1940, Moscow Univ. Uchenye zapiski, Wissensch. Ber. 42:85.

Cheyletus intrepidus Oudemans, 1903*b*, Ent. Ber. Nederl. Ent. Ver. 1(12):84. Male.

Cheyletus longipalpus Ewing, 1909 (see under *Paracheyletia pyriformis*).

Cheyletus longipes Mégnin, 1878 (see under *Cheletomorpha lepidopterorum*).

Cheyletus nigripes[3] Mola, 1907, Zool. Anz. 32:43.

Cheyletus parumsetosus Karpelles, 1884, Berl. Entom. Zeitschr. 28(2):231–244.

Cheyletus patagiatus[3] Nordenskiöld, 1900, Medd. Soc. Faun. Fenn. 26:37.

Cheyletus polymorphus Volgin, 1949, C. R. Acad. Sci. U.R.S.S. Moscow, N. S. 64:584.

Cheyletus praedibundus Rohdendorf (not Kuzin), 1940, Moscow Univ. Uchenye zapiski, Wissensch. Ber. 42:85.

Cheyletus promptus Oudemans, 1904*b*, Ent. Ber. Nederl. Ent. Ver. 1(18):161.

Cheyletus pyriformis Banks, 1904 (see under *Paracheyletia*).

Cheyletus rapax Oudemans, 1903*b*, Ent. Ber. Nederl. Ent. Ver. 1(12):84. Male.

Cheyletus rohdendorfi Zachvatkin, 1949; *see under* Volgin, 1955, Akad. Nauk S.S.S.R., Zool. Inst., Opredel. p. Faune S.S.S.R. no. 59:159.

Cheyletus saevus[3] Oudemans, 1904*b*, Ent. Ber. Nederl. Ent. Ver. 1(18):161.

Cheyletus schneideri Oudemans, 1904*c*, Tijdschr. Nederl. Dierk. Ver. (ser. 2) 8:Versl. Xv.

Cheyletus strenuus Oudemans, 1904*b*, Ent. Ber. Nederl. Ent. Ver. 1(18):161.

---

[3] Listed as "Uncertain Species" by Baker, 1949.

Cheyletus truculentus Volgin, 1949, C. R. Acad. Sci. U.R.S.S. Moscow, N. S., 64:586.
Cheyletus trux Rohdendorf, 1940, Moscow Univ. Uchenye zapiski, Wissensch. Ber. 42:87.
Cheyletus ugandanus Lawrence, 1954, Ann. Natal Mus. 13(1):65–67.
Cheyletus venator Vitzthum, 1920 (*Cheletes venator*), Arch. Naturg. 84A (6):2.
Cheyletus vorax Oudemans, 1903*b*, Ent. Ber. Nederl. Ent. Ver. 1(12):84.

Eucheyletia Baker

Eucheyletia Baker, 1949, Proc. U.S. Nat. Mus. 99(3238):294.
Cheyletia (in part), A. M. Hughes, 1961, Ministry Agric., Fish. and Food Tech. Bull. 9:188.
Zachvatkiniola Volgin, 1969, Akad. Nauk S.S.S.R., Zool. Inst., Opredel. p. Faune S.S.S.R. no.
101:156. Type: *Eucheyletia reticulata* Cunliffe. New synonym.

Palp tarsus with 2 sickles, 2 combs. Palp claw with 2 to 4 basal teeth. Protegmen conelike, drawn into apical process which merges with rostrum. No eyes. Idiosoma entirely or mostly covered by 2 large dorsal plates. Dorsolateral body setae fanlike. Dorsomedians aberrant: squamate, amoeboid or cloudlike; 6 to 15 pairs. With 1 pair of anal setae fanlike (except *hardyi*). Tarsi I–IV with claws and rayed empodia.

Type species.—Eucheyletia bishoppi Baker; by original designation.

KEY TO FEMALES OF EUCHEYLETIA
(Adapted from Volgin, 1963*a*)

1. Palp claw with 2 basal apophyses 2
   Palp claw with 3 basal apophyses 11
2. All anal setae acicular; hysterosomal plate shorter than propodosomal plate, bears only
   2 pairs of leaflike marginal setae ....*hardyi* Baker
   Posteriormost pair of anal setae leaflike; hysterosomal plate not shorter than propodosomal plate 3
3. With 1st pair of marginal setae on hysterosoma set on trivial platelets separated from
   principal median plate 4
   With 1st pair of marginal setae on hysterosoma set on principal median plate 9
4. Ventral seta of palp genu acicular, smooth or nearly so 5
   Ventral seta of palp genu spatulate or like a narrow scale 8
5. Tibia I about as long as tarsus I; length of tibia I exceeds 5 times its greatest diameter
   sinensis Volgin
   Tibia I about two-thirds as long as tarsus I; length of tibia I exceeds 3 times its greatest
   diameter 6
6. Hysterosomal plate bears 3 pairs of leaflike marginal setae; dorsomedian setae absent
   flabellifera (Michael)
   Hysterosomal plate bears 5 pairs of leaflike marginal setae; 7 or 8 pairs of peculiarly
   modified median setae on each main dorsal plate 7
7. Guard seta of tarsus I lightly barbed, longer, somewhat thicker than solenidion *w*I;
   ventral seta on palp genu lightly barbed; hysterosomal plate bears 7 pairs of modified
   median setae ....*bothrophila* Volgin
   Guard seta of tarsus I smooth, considerably shorter than *w*I; ventral seta on palp genu
   smooth; hysterosomal plate bears 8 pairs of modified median setae ....*taurica* Volgin
8. Guard seta on tarsus I thin, acicular, minute ....*eoa* Volgin
   Guard seta on tarsus I lanceolate, coarsely barbed, longer than *w*I ....*pavlovskyi* Volgin
9. Dorsal seta of palp tibia leaflike; 5 pairs of marginal setae on hysterosomal plate;
   median setae absent on both main plates ....*womersleyi* Volgin
   Dorsal seta of palp tibia acicular, smooth; 6 pairs of marginal setae on hysterosomal
   plate; each median dorsal plate bears 7 or 8 pairs of cloudlike median setae 10
10. Each median dorsal plate with 7 pairs of cloudlike median setae ....*bishoppi* Baker
    Each median dorsal plate with 8 pairs of cloudlike median setae ....*asiatica* Volgin

11. Propodosomal plate with 5 pairs of setae; dorsomedian setae similar to marginals; dorsal plating coarsely reticulate; tibia I with 4 "tactile" setae—2 lanceolate or leaflike, 2 acicular ....*reticulata* Cunliffe

Propodosomal plate with 7 to 11 pairs of setae; dorsomedian setae peculiarly modified; dorsal plates not coarsely reticulate; tibia I with 5 "tactile" setae—4 lanceolate or leaflike, 1 acicular 12

12. Hysterosomal plate bears 6 pairs of marginal setae; propodosomal plate bears 7 pairs of median setae ....*sibirica* Volgin

Hysterosomal plate bears 5 pairs of marginal setae; propodosomal plate bears 3 to 5 pairs of median setae 13

13. Hysterosomal plate 1.25 times as long as propodosomal plate; 3 pairs of median setae on each main plate; guard seta of tarsus I lanceolate, pubescent, twice as long as wI
harpyia (Rohdendorf)

Median dorsal plates about equal in length; 5 pairs of median setae on propodosomal plate, 6 pairs on hysterosomal plate; guard seta of tarsus I acicular, barbed, 3 times as long as wI ....*bakeri* Volgin

Eucheyletia bishoppi Baker
(Fig. 24)

Eucheyletia bishoppi Baker, 1949, Proc. U.S. Nat. Mus. 99(3238):295–296.

A relatively large cheyletid. Palp claw with 2 basal teeth, axis of proximal tooth inclined toward tip of distal tooth. Outer comb ca. 15 teeth, several distal teeth very much longer than those of median length; terminal extension of comb scimitar-like, tip rounded. Inner comb ca. 30 nearly isometric teeth. Ventral seta on palp genu spatulate, barbed and ribbed. Protegmen extensive, subconic, merged—i.e., lengthened to a point where this part of stylophore presses into gutter of rostrum. Tegmen with 6 to 8 longitudinal ridges of irregular widths; ridges bifurcate and anastomose to form a coarse reticulum; small, ill-defined or flocculate tubercles interspersed between raised ridges; protegmen similarly sculptured. Peritremes implanted in heavily sclerotized groove, obtusely bent rearward, 3 links in transverse arms, 3 or 4 links in descending arms. Dorsal setae: 10 pairs of conventional dorsolaterals, 12 to 15 pairs of aberrant dorsomedians. Dorsolaterals (and humerals) broadly rounded, spatulate (31–66), perimeter of blade with outline similar to that of a hanging drop of fluid, 7 to 9 barbed ribs on convex surface, an equal or greater number of thin-lined ribs on concave surface; membrane delicately net-veined between ribs; 4 pairs on propodosomal plate, at least 5 pairs on hysterosomal plate. Dorsomedian setae with noticeably long basal stalks; distal parts of each seta flattened and branched into an amorphous mass or group of pseudopod-like fingers which intermingle with those of adjacent setae; 6 or 7 pairs on propodosoma, 6 to 8 pairs on hysterosoma. Dorsal plating densely pebbled or tuberculate, tubercles with hazy outlines; spacing between tubercles about one diameter of a tubercle; arrangement of tubercles in rows not very obvious. Anterior margin of propodosomal plate concave, heavily sclerotized, sharply outlined. Body striae double-lined, tuberculate on dorsum and ventral opisthosoma, plain on intercoxal integument. Posterior-most pair of anal setae spatulate.

Length ratio: leg I/idiosoma = 0.7. Setae on podomeres: femora 2-2-2-2, genua 3-2-2-2, tibiae 6-4-4-4, tarsi 10-8-7-7. Anterior seta on coxa III spatulate; all other coxal setae acicular, smooth. Tarsus I: addorsals *tc* rough or minutely annulate; solenidion wI moderately long (26); guard seta (57) acicular, very sparsely barbed, twice as long as wI; ventral seta v thick, abundantly barbed. Measurement (n = 10): length idiosoma 454±24, gnathosoma 175±13, leg I 304±9, tarsus I 119±7.

Collection data.—California: soil sample, Winters, Yolo Co. (L. M. Smith and R. O. Schuster); manzanita leaf mold, Oakville, Napa Co. (Irwin and Cavagnaro); *Neotoma* nest, Oakville (I. and C.); topsoil beneath Bishop pine, Tomales Bay State Park, Marin Co. (G. A. Marsh); topsoil, Cruickshank Estate, Pebble Beach, Monterey Co. (L. M. S.); topsoil, Belmont, San Mateo Co. (R. O. S.); leaf mold, Kaiser Pass, Fresno Co. (R. O. S.); *Neotoma* nest, Modjeska Canyon (Irwin and Bath).

Eucheyletia sinensis Volgin

(Fig. 25)

Eucheyletia sinensis Volgin, 1963*a*, Akad. Nauk S.S.S.R., Zool. Inst., Parazitol. Sbornik 21:50–52.

Palp claw with 2 blunt teeth in a kind of thumb-forefinger arrangement. Outer comb ca. 17 teeth, terminal tooth scimitar-like; inner comb > 30 teeth. Ventral seta of palp genu acicular, with few incipient barbs. Dorsal seta of palp femur appears frayed or truncate at end, possibly owing to normal wear or breakage in mounting. Rostrum spade-shaped, conspicuously displayed; rostrum and constricted stylophore give mouthparts a duckbill appearance. Protegmen narrow, barely capacious enough to enclose basal sclerites of stylets, apex pointed and laid in rostral gutter; junction of tegmen and protegmen marked by a deep transverse groove; crossarms of peritremes set in posterior wall of this groove. Dorsal cuticle of tegmen abundantly stippled with small tubercles. Dorsal body plating very lightly sclerotized, with small granules or micro-tubercles arranged in rows. Interscutal striae also finely granulate. Hysterosomal plate confined to area bounded by 2d to 6th pairs of dorsolateral setae; 1st pair of dorsolateral hysterosomals set on independent platelets in front of main shield. Dorsal body setae: 27 pairs plus humerals. Dorsolateral setae: 10 pairs (35–55); humerals slightly longer (59–70); blades oval in outline, with 7 or 8 rows of uniform blunt spinelets. Dorsomedian setae squamate, overlapping; support-ing framework of each seta appears as a fine-meshed flocculent reticulum; 9 pairs of propodo-somal plate, 8 pairs on hysterosomal plate. Posterior anal seta fan-shaped, about as large as hind-most dorsolaterals.

Length ratio: leg I/idiosoma = 0.8. Setae on legs I to IV: femora 2-2-2-2, genua 3-2-2-2, tibiae 6-4-4-4, tarsi 10-8-7-7. Inner border of coxa II bears a rounded apophysis. Sternal integument has a pair of small sclerites. Tarsus I has a short solenidion wI (30) and a minute guard seta in duplex position. Measurements (n = 3); length idiosoma 517, gnathosoma 200, leg I 390, tarsus I 120.

Collection data.—Illustrated specimen taken from *Eothemonys melanogaster*, Sung Kong, Nan Tou Hsien, Formosa. Volgin's type material of *sinensis* taken from *Rattus humiliatus celsus* Gl. Allen, Hunnan Province, Chinese People's Republic. A specimen deposited at UCD is from a rodent, Nepal (Phulung Ghyang Newakot Dist., 11,000 ft.).

Eucheyletia hardyi Baker

(Fig. 26)

Eucheyletia hardyi Baker, 1949, Proc. U.S. Nat. Mus. 99(3238):196–297.

This species so closely resembles *bishoppi* in size and conformation that only its contrasting features need emphasis here.

Outer comb ca. 18 teeth; inner comb ca. 35 teeth. Ventral seta on palp genu acicular, smooth. Hysterosomal plate covers approximate tergal area of metapodosoma, bears only 2d and 3d pairs of dorsolateral hysterosomal setae. Dorsal body setae 15 or 16 pairs plus humerals. Dorsolaterals have a short size range (50–59). Dorsomedian setae cloudlike or flattened squames having vari-able outlines; each seta comprises tangled stands of beadlike tubercles fixed in a transparent matrix; squamous blade may be notched to accommodate a very short stalk; 2 or 3 pairs on propodosomal plate, 3 pairs on hysterosomal plate. Posteriormost anal setae acicular, smooth. Femur IV has only 1 seta. Measurements (n = 1): length idiosoma 430, gnathosoma 176, tarsus I 117, solenidion wI 17.

Collection data.—Illustrated holotype specimen, USNM No. 1769, taken from a nest of *Neotoma micropus*, Harlingen, Texas; specimen mounted in a resinous medium and slightly distorted by shrinkage.

Eucheyletia reticulata Cunliffe
(Fig. 27)

Eucheyletia reticulata Cunliffe, 1962, Proc. Ent. Soc. Wash. 64(3) :200–201.
Eucheyletia reticulata Volgin, 1963a, Akad. Nauk S.S.S.R., Zool. Inst., Parazitol. Sbornik
 21:63–65.
Zachvatkiniola reticulata (Cunliffe), Volgin, 1969, Akad. Nauk S.S.S.R., Zool. Inst., Opredel.
 p. Faune U.S.S.R. No. 101:157–158.

Palp claw with 3 or 4 short splayed teeth. Outer comb ca. 15 teeth; inner comb > 35 teeth.
Ventrolateral seta of palp genu fanlike, resembling dorsal seta of this segment. Dorsal palp
femoral seta fan-shaped but skewed to a blunt point on mesal tip. Inflated part of protegmen
comprises 0.4 of stylophore length. General body plating dimpled, with pits crowded together to
produce a surface reticulation of cuticle; reticulum appears on tegmen, basis capituli, palp
femora, coxae I–IV, humeral plates, and 2 major tergal sclerites. Humeral plates are rather large
oval discs (35 on long dia.), displaced to ventrolateral position. Peritremes small in caliber,
4 links per side. Dorsal body setae broadly spatulate or fan-shaped, fluted and delicately barbed,
ends rounded; 14 pairs plus humerals, all similar in form. Propodosomal plate bears 1 pair of
dorsomedians; hysterosomal plate has 3 pairs of dorsomedians plus 1st to 5th pairs of dorsolaterals (inclusive). Posterior anal setae spatulate.

Length ratio: leg I/idiosoma = 0.7. Setae on legs I to IV: femora 2-2-2-1, genua 3-2-2-2, tibiae
5-4-4-4, tarsi 9-8-7-7. Tarsus I with conspicuously long (62) lanceolate guard seta on its own
tubercle behind shorter solenidion wI (27). Measurements (n = 1): length idiosoma 312, gnathosoma 101, leg I 220, tarsus I 94.

Collection data.—Two study specimens taken from *Oryza sativa*, Japan, intercepted at Hawaii.

Volgin (1969) erected *Zachvatkiniola* for *Eucheyletia reticulata* Cunliffe, 1962.
We believe that the reticulation of the skeleton, although very pronounced in this
species, is a specific rather than a generic character. The reticulation of the skeleton
and the fact that the four pairs of dorsomedian setae are orthodox in females are
good spot characters for recognizing the species. Otherwise, however, the species
appears to be congeneric with the type of *Eucheyletia*. The taxonomy of this group
of species will be simpler if *reticulata* is retained in *Eucheyletia*.

The genus *Eucheyletia* contains fourteen species all of which were included
in Volgin's (1963a) review of the genus. Only three of these species—*bishoppi,
hardyi, bakeri*—are known to occur in North America. The species for which
examples could not be obtained for this study are listed below.

Eucheyletia bothrophilia Volgin, 1963a, Akad. Nauk S.S.S.R., Zool. Inst., Parazitol. Sbornik
 21:44–47. Found in nests of various small rodents, Trans-Carpathian Province.
Eucheyletia flabellifera (Michael)
 Cheyletia flabellifer Michael, 1878, Jour. Roy. Micr. Soc. 1:135–138.

The structural peculiarities of this classical species are not sufficiently described
by Michael (1878) and Oudemans (1904, 1906) to provide assured identifications.
The mite which A. M. Hughes (1948) illustrated as *flabellifera* (Michael) has
been renamed *taurica* Volgin, and those which Womersley identified as *flabellifera*
(Michael) have been renamed *womersleyi* Volgin (Volgin, 1963a). Uncertainty
about the recognitional characters of Michael's species may be dispelled by a
thorough study of the *Eucheyletia* species found in the type locality—Yorkshire
County, England, in cellar dust.

Eucheyletia taurica Volgin, 1963a, pp. 48–50.
 Cheyletia flabellifera (Michael) of A. M. Hughes, 1948, 1961.

Redescribed by Volgin from specimens captured in nests of a common field mouse (*Microtus arvalis* Pall.) and the forest mouse (*Apodemus sylvaticus* L.), territory of the Crimean Reservation, near Gursuf. Volgin equates these mites with those which Hughes called *flabellifera* (Michael).

Eucheyletia womersleyi Volgin, 1963a, p. 59.
   Cheyletia flabellifera (Michael) of Womersley, 1941, Rec. So. Australian Mus. 7(1):60.

The Australian material was found in the debris of an old Yacca (*Xanthor-rhoea*) stump in Torrens Gorge, South Australia. Womersley believed that rabbits probably nested nearby.

Eucheyletia eoa Volgin, 1963a, pp. 52–54. Type specimens taken from nest of a Siberian chip-
   munk, *Eutamias sibirica* Laxm., Olginsk Region, Maritim Territory; specimens also found
   on the forest mouse, *Apodemus speciosus* Pemmink, Kalinin Region, Maritim Territory.
Eucheyletia pavlovskyi Volgin, 1963a, pp. 54–56. Found in a nest of the Steppe "petrushka,"
   Lagurus lagurus Pall., Karaganda Oblast, Kazakh S.S.R.
Eucheyletia asiatica Volgin, 1955, Akad. Nauk S.S.S.R., Zool. Inst., Opredel. p. Faune S.S.S.R.
   no. 59:166; also Volgin, 1963a, pp. 58–59. From nests of various small rodents, provinces of
   Alma-Ata, Trans-Carpathia, Karaganda.
Eucheyletia harpyia (Rohdendorf)
   Cheyletia harpyia Rohdendorf, 1940, Moscow Univ. Uchenye zapiski, Wissensch. Ber. 42:90–92.
   Type locality: Ivanovskay Oblast, in a granary.

The male was redescribed by Volgin (1963a) from one of the type specimens in the collection of the Zoological Institute, Academy of Sciences U.S.S.R. According to Volgin, the female specimen (paratype?) described briefly by Rohdendorf has been lost.

Eucheyletia sibirica Volgin, 1963a, pp. 61–62. Females of type series found in oat seeds, Asinovsk
   Region, Tomsk Province (Ratanova).
Eucheyletia bakeri Volgin, 1963a, pp. 62–63.
   Eucheyletia harpyia (Rohdendorf) of Baker, 1949, Proc. U.S. Nat. Mus. 99(3238):294–295.
   Holotype (female) from nest of a bumblebee, *Bombus* sp., Beaver Mountain, Alaska. Slide
or slides containing type specimens were not found among cheyletids on file in USNM.

Cheletophanes Oudemans

Cheletophanes Oudemans, 1904b, Ent. Ber. Nederl. Ent. Ver. 1 (18):162.

This genus is known to us only through Oudemans' (1906) redescription of *Cheletophanes montandoni* (Berlese and Trouessart).

A large species (680–800), with leg I relatively long. Palp claw bears ca. 13 teeth distributed over its entire length. Palp tarsus with 2 combs and 2 sickles. Two plates entirely cover dorsum of idiosoma, both sculptured to show several ridges or concentric striae encircling some dorso-median setae. Eyes present. Dorsal body setae slender spatulate, profusely pilose, most just about long enough to overlap bases of those next behind; laterals and medians similar in form. Oude-mans locates 8 pairs of setae on propodosomal plate (4 laterals plus 4 medians), and 9 or 10 pairs on superior part of hysterosomal plate; this plate is reflected onto ventrolateral parts of opisthosoma and carries an additional 3 pairs on its inferior extension. The genus appears to be very unusual in having superfluous setae on venter of opisthosoma: 4 pairs resembling dorsal setae, arranged in a crescentic row across anal area, plus 2 or 3 pairs of short anal setae. Paired claws and empodia on all tarsi.

Type species.—Cheyletus montandoni Berlese and Trouessart, 1889, Bull. Biblio. Scient. Ouest II, 2(9):133; by original designation. From beneath elytra of *Aradus varius*, Brostenii (Valachie du nord).

The only other species assigned to this genus is *Cheletophanes peregrinus* Berlese, 1921, Redia 14:194.

Cheletacarus Volgin

Cheletacarus Volgin, 1961, Akad. Nauk S.S.S.R., Zool. Inst., Parazitol. Sbornik 20:248.

Palp claw toothed along greater part of mesal margin. Palp tarsus bears 2 combs and 2 sickles. Protegmen elevated, extending far in front of peritremes, covering rostrum to origins of superior adoral setae. Peritremes shaped like inverted U; rearwardly directed arms comprise 9 short overlapping links. Dorsal body plating consists of one large thin shield on propodosoma and numerous trivial platelets, one for each seta not implanted on single major propodosomal plate. One pair of inconspicuous eyes on margins of shield. Dorsal body setae spatulate or conservatively fan-shaped in females, all similar in form and relatively small. All tarsi with claws. Solenidion *w*I very long, with no obvious guard seta.

Type species.—Cheletacarus raptor Volgin, 1961; by original designation.

Cheletacarus gryphus, new species
(Fig. 28)

Palp claw with 12 to 14 teeth distributed on mesal surface. Outer comb ca. 20 teeth, distal tip terminating in flat recurved blade; inner comb ca. 25 teeth. Tegmen covered almost to basal crossbrace with closely set rows of elliptical tubercles. Protegmen with somewhat smaller tubercles in more widely spaced rows. Tuberculate striae also cover basis capituli and completely enwrap palp femora. Propodosomal shield weakly sclerotized, surface ornamented with small granules or tubercles disposed in sinuous rows or whorls. Striae on unplated dorsum appear as linear strands of tiny, equispaced rodlike elevations ("dotted" striae). Ventral striae plain. Eyes present. Dorsal body setae short (27–43) and narrow, wedge-shaped with almost truncate ends, 4 or 5 finely barbed ribs; 17 pairs plus humerals. Median setae resemble laterals, 3 pairs on propodosomal plate, 4 pairs on hysterosoma. Setae on venter flagelliform, smooth; posterior seta on coxa I long enough to reach across ventral midline. Paragenital setae (2 pairs) unusually long, about 4 times as long as genitals (2 pairs on flaps).

Length ratio: leg I/idiosoma = 0.7. Setae on legs I to IV: femora 2-2-2-1, genua 3-2-2-2, tibiae 6-5-4-4, tarsi 9-8-8-7; each tibia II to IV has ultralong flagelliform seta in addorsal position; on tibia IV this seta is at least 125; extra seta on tarsus III is a tiny solenidion. Tarsus I: solenidion *w*I (82) about as long as tarsus less pedicel; guard seta seems to be absent; claws of tarsus I (8) much shorter and less robust than those of tarsus II (18). Measurements (holotype): length idiosoma 512, gnathosoma 172, leg I 372, tarsus I 125.

*Holotype.—*Female, from bark of almond tree, Davis, Calif. (D. W. Price). Sole specimen; deposited in UCD.

This species appears to be most closely allied to species of *Cheletomimus*, especially *C. berlesei.* It also has affinities with *Hemicheyletia* as well as *Cheletonella.*

Cheletacarus raptor Volgin

Cheletacarus raptor Volgin, 1961, Akad. Nauk S.S.S.R., Zool. Inst., Parazitol. Sbornik 20:248–253.

According to Volgin, the females of *raptor* have two pairs of paragenital setae which are not appreciably longer than the setae of the two pairs of genitals on the covers of this orifice. In *gryphus* the two pairs of paragenital setae are very long and flagelliform; each seta is at least three times as long as any one of the four setae (two pairs) on the genital covers. It is possible also that the minute solenidion observed on tarsus III of *gryphus* (female) may be peculiar to that species.

Cheletacarus rugosus (Womersley)

Cheletophanes rugosa Womersley, 1941, Rec. So. Australian Mus. 7 (1):62–63.
Cheletacarus rugosus (Womersley); Volgin, 1961.

Volgin's assignment of Womersley's species to *Cheletacarus* rests on the premise that the female of the species lacks a median hysterosomal shield. Although Womersley illustrated both sexes of this mite, his study of the female was based upon one damaged specimen, presumably a syntype. He presented as much detail as allowed by this imperfect specimen. The detail which is missing on the female of the type series may be crucial to an understanding of its systematics. The placement of the species in this genus is a reasonable option according to the information available, but is should be regarded as provisional until new material is collected.

Cheletonella Womersley

Cheletonella Womersley, 1941, Rec. So. Australian Mus. 7(1):60; Baker, 1949, Proc. U.S. Nat. Mus. 99(3238):292; Volgin, 1955, Akad. Nauk S.S.S.R., Zool., Inst., Opredel. p. Faune S.S.S.R. no. 59:166–168.

Relatively large, soft-bodied cheyletids having a pointed beak. Palp claw with few basal teeth. Palp tarsus with 2 sickle-like and 2 comblike setae on stylophore approximately triangular, its protegmenal portion conical and almost entirely covering free end of rostrum. No eyes. Body plating restricted to a single shield on propodosoma; dorsal setae not inserted on this shield are borne on trivial platelets, one for each seta. Dorsal body setae conservatively fan-shaped, longer than wide, none aberrant or peculiarly fashioned. Legs shorter than idiosoma. Tarsus I bears a moderately long solenidion wI set a short distance in front of a longer guard seta.

Type species.—Cheletonella vespertilionis Womersley. Monotypic.

Cheletonella vespertilionis
(Fig. 29)

Cheletonella vespertilionis Womersley, 1941, Rec. So. Australian Mus. 7(1):61.

Palp tarsus bears 3 basal teeth, middle tooth largest. Outer comb ca. 16 teeth; inner comb ca. 20 teeth. Palp tarsus with 2 normal sickle-like setae. Palp tibia wedge-shaped, with a strong, straight-edged flange overhanging tarsal joint; dorsal palptibial seta acicular, smooth. Dorsal setae on palp genu and femur resemble those on propodosoma. Stylophore almost triangular in outline. Protegmen conical, covering rostrum to tip, about half as long as tegmen, not ornamented. Tegmen roughened with thick, broken striae; small pores scattered over its total area. Peritremes with short transverse arms comprised of 2 or 3 links, posterior arms of about 7 links; peritremes circumscribe front and sidewalls of tegmen almost to its posterior crossbrace. Rostrum pointed, not projecting forward beyond tip of stylophore. Idiosoma widest where humeral setae arise. A single rectangular plate covers midpart of propodosoma; no plates on hysterosoma other than trivial platelets on which individual setae arise; propodosomal plate sculptured like stylophore. Dorsal body setae short (35–59); blades transparent, conservatively widened fanwise to approximately 30° sectors of a circle, with 4 to 7 lightly barbed ribs; 14 pairs plus humerals; dorsolaterals (10 pairs) and dorsomedians (4 pairs) similar in form. Propodosomal plate bears 3 pairs of dorsolaterals on its front and side margins, 1 pair of dorsomedians on posterior margin. Anal protuberance with 3 pairs of anal setae in a semicircle, with middle and hindmost pairs symmetrically forked (bifid).

Length ratio: leg I/idiosoma = 0.7. Setae on legs I to IV: femora 2-2-2-1, genua 3-2-2-2, tibiae 5-4-4-4, tarsi 10-8-7-7. Tarsus I: azygos seta in ventrolateral position, considerably behind mesal infraterminal; solenidion wI moderately long (30); guard seta longer than wI, acicular, very sparsely barbed. Measurements (n = 3): length idiosoma 538, gnathosoma 191, leg I 395, tarsus I 133.

*Collection data.—*California: from bat guano, 3 miles S of Calistoga, Napa Co. (A. Beck); from *Pinus aristata* (soil), Crooked Creek Research Sta., Mono Co. (W. H. Sholes).

The few specimens collected from bat guano and forest soil in California appear to be conspecific with the holotype specimen taken from a bat in Australia. The type was loaned to us by the South Australian Museum, Adelaide.

Another species has been described in *Cheletonella* by Volgin:

Cheletonella caucasica Volgin, 1955, Akad. Nauk S.S.S.R., Zool. Inst., Opredel. p. Faune S.S.S.R. no. 59:168.

According to Volgin (1969), the one pair of dorsomedian setae on the propodosomal plate of *caucasica* is inserted noticeably behind the fourth dorsolaterals on the integument beside the propodosomal plate; all the anal setae are acicular, smooth, not forked. In *vespertilionis* the fourth pair of dorsolateral propodosomal setae is approximately aligned in a straight crossrow with the single pair of dorsomedians on this part of the body; two of three pairs of setae on the anal swelling are forked close to their tips (fig. 29*a*).

Cheletonata Womersley

Cheletonata Womersley, 1955, Proc. Linn. Soc. N. S. Wales 80(3) :215.

Palp claw with basal teeth. Palp tarsus bears 2 sickles and 1 robust comb; homologue of inner comb is a smooth acicular seta. Stylophore ornamented with broken striae longitudinally disposed, and 8 to 10 coarse ridges forming a crude reticulum on its forward half; protegmen subconic, gently elevated over stylet bases, then tapering to pointed processes fitted into rostral gutter. Rostrum pointed, projecting considerably beyond protegmen. Peritremes M-shaped. Propodosoma partly covered by a median plate. Eyes present on anterolateral corners of propodosomal plate, between 2d and 3d marginal setae. Hysterosoma bears one small median plate; no other sclerites on hysterosoma except tiny platelets associated with individual setae. Dorsal body setae spatulate or conservatively fan-shaped or acicular; no structural difference between medians and laterals. Legs slender, 1st and 4th shorter than idiosoma. Tarsi I to IV bear smooth claws and rayed empodia; claws on tarsus I minute. Solenidion *w*I on tarsus I just less than half as long as entire segment; guard hair a microseta, in duplex arrangements with *w*I.

Type species.—*Cheletonata milesi* Womersley, 1955; by original designation.

Cheletonata milesi Womersley
(Fig. 30)

Cheletonata milesi Womersley, 1955, Proc. Linn. Soc. N. S. Wales 80(3) :215–216.

Basal process on palp claw bears 3 to 5 short conical teeth. Outer sickle longer than claw. Outer comb with sturdy shaft, ca. 14 bluntly pointed teeth. Homologue of inner comb is a slender flagelliform seta having no barbs. Integument of stylophore provided with broken striae longitudinal indirection and 8 to 10 elevated wrinkle-like ridges which form an erratic meshwork on protegmen and anterior part of tegmen; protegmen subconic, elevated over stylet bases then tapering to pointed processes fitted into rostral gutter. Rostrum pointed, projecting considerably beyond distended part of protegmen. Peritremes with 7 to 9 slender links per side; transverse arms turn rearward through an arc of short radius, not bent into an acute angle. Idiosoma ovoid, partly plated on dorsum. Propodosomal plate quadrangular, widest across front margin; encompasses all median setae but only front 3 pairs of laterals, faintly sculptured over-all; central area has fine network of trabeculae which becomes a fairly coarse reticulum along both side margins. Median hysterosomal plate small, oval in outline on midpart of metapodosoma between 1st and 2d pairs of median setae, with a delicate reticulation but no setae. Dorsal body setae spatulate or narrow fans having straight sides and rounded ends; blades thin, transparent, with about 5 barbed ribs on upper surfaces; median and lateral setae identical in form and relatively small size range (39–59) between setae on different areas of dorsum; altogether 16 pairs plus humerals; propodosomal plate bears 3 pairs of laterals plus 3 pairs of median; 4th laterals on independent trivial platelets; hysterosoma has 6 pairs of laterals plus 3 pairs medians, each on a separate trivial platelet; last pair of laterals on venter. Paragenitals 2 pairs, genitals 2 pairs, anals 3 pairs with 1 pair barbed but not spatulate.

Length ratio: leg I/idiosoma = 0.6. Setae on legs I to IV: femora 2-2-2-1, genua 3-2-2-2, tibiae 6-4-4-4, tarsi 10-8-7-7. Tarsus I: azygos seta arises far back, behind addorsals *tc* by one-third distance *tc* to *w*I; solenidion *w*I slender, pointed, long enough (61) to reach alveoli of *tc*; claws very small. Measurements (n = 4): length idiosoma 656, gnathosoma 210, leg I 423, tarsus I 148.

Collection data.—Type specimens collected from abandoned nest of the wren, *Malurus melanocephala*, in Northern Australia.

The foregoing redescription is based upon a short series of type specimens loaned by the South Australian Museum, Adelaide. Our measurements do not agree with Womersley's because different reference points were used. The small median hysterosomal plate was not indicated in the original description, and Womersley seems to have misidentified the eyes. We concur with Womersley's opinion that the genus is closely allied to Cheletonella.

Paracaropsis Volgin

Paracaropsis Volgin, 1969, Akad. Nauk S.S.S.R., Zool. Inst., Opredel. p. Faune S.S.S.R. no. 101:322–323.

Palp claw with 8 or 9 teeth. Homologue of inner comb appears to be a small acicular, smooth sensillum. Rostrum conspicuously extended forward from beneath distended part of protegmen, its sides almost parallel, tip rounded or spadelike. Eyes present. Hysterosomal plating restricted to 2 small unpaired plates; 1 plate on metapodosoma, which bears no setae, between 1st and 2d pairs of median hysterosomal setae; 2d plate in suranal position bears last 3 pairs of dorsal setae. Dorsal body setae short, acicular, smooth; humeral seta ultralong, smooth, flagelliform. Dorsomedian and dorsolateral setae alike in form.

Type species.—*Acaropsis strofi* Samšiňák, 1956; by original designation. It now appears that *strofi* is a synonym of *Acaropsis travisi* Baker, 1949.

The assignment of *travisi* to a separate genus, as advocated by Volgin, eliminates some of the complications in the concept of the parent genus, *Acaropsis*. The general organization of *Paracaropsis travisi* (=*strofi* of Samšiňák) seems to show as much resemblance to *Cheletonata milesi* as to the best-known species of *Acaropsis*.

Paracaropsis travisi (Baker)
(Fig. 31)

Acaropsis travisi Baker, 1949, Proc. U.S. Nat. Mus. 99(3238):313–314.

Paracaropsis travisi (Baker); Volgin, 1969, Akad. Nauk S.S.S.R., Zool. Inst., Opredel. p. Faune S.S.S.R. no. 101:322–325.

Paracaropsis strofi Samšiňák, 1956. New synonym.

Palp claw bears 8 or 9 teeth. Palp tarsus with 1 comb and 2 sickles; comb ca. 14 teeth; homologue of inner comb appears to be a small acicular sensory element. Other setae of palp segments slender, flagelliform, smooth. Rostrum narrow, bullet-shaped at tip, lateral margins almost parallel behind adoral setae; about half as long as midline length of gnathosoma. Protegmen subconic; its margins converge and taper to a long, pointed process which presses into median gutter of rostrum. Each transverse arm of peritremes obtusely bent at 3d link; 4 or 5 links in each descending arm. Cuticula of gnathosoma lightly sclerotized, with a few small scattered pores but no distinctive surface ornamentation except for a small patch of sharply raised, broken striae on each maxillicoxa, opposite last 2 links of peritremes; anteriormost margin of tegmen shows a fringe of short longitudinal striae close by transverse arms of peritremes. Idiosoma slightly rhomboid or lozenge-shaped, with 3 lightly sclerotized dorsal plates; anterior plate bears only 5 pairs of setae; posteriormost dorsolateral setae of propodosoma not inserted on this plate. Eyes protuberant, with 4 concentric striae around each cornea. Hysterosoma bears 2 median plates:

a small anterior plate in an area bounded by 1st and 2d pairs of dorsomedian setae (these setae not on plate) and a small suranal plate on which are borne last 3 pairs of dorsal setae. Dorsal body setae tend to be short (16–24), acicular, bluntly pointed, not obviously barbed, many set on individual, trivial platelets; total number 14 pairs plus humerals. Humeral setae ultralong (at least 120), flagelliform. Paragenitals 2 pairs, genitals 2 pairs, anals 3 pairs, all acicular, smooth.

Length ratio: leg I/idiosoma = 0.8. Setae on legs I to IV: femora 2-2-2-1, genua 3-2-2-2, tibiae 5-5-4-4, tarsi 10-8-7-7. Lateral dorsal setae on tibiae II to IV at least 80, with each seta longer than any ventral seta on either of these podomeres. Solenidion wI (35) about half as long as body of tarsus I; a minute guard seta arises close behind wI. Smooth claws and multirayed empodia on all tarsi. Measurements (holotype only): length idiosoma 266, gnathosoma 113, leg I 205, tarsus I 82.

This redescription of *P. travisi* is based on a study of the holotype, USNM 1776, for a lizard, *Sceloporus woodi* (?), Newton, Georgia.

Cheletophyes Oudemans

Cheletophyes Oudemans, 1914, Ent. Ber. Nederl. Ent. Ver. 4(78):101; Baker, 1949, Proc. U.S. Nat. Mus. 99(3238):287–288; Volgin, 1965, Akad. Nauk S.S.S.R., Trudy Zool. Inst. 35: 289–290.

Palp claw with 3 teeth; palptarsal combs with few coarse teeth, 6 to 9 teeth. Mouthparts well tanned; rostrum narrow, apically pointed; protegmen domed, with convergent margins, apex depressed. Peritremes M-shaped, large in caliber, links short, bulbous. Idiosoma with 2 median dorsal plates and 8 or 9 pairs of trivial platelets; anterior median plate covers most of propodosoma; posterior median plate small, confined to suranal part of opisthosoma. Eyes present. Dorsal setae rodlike, relatively short, coarsely barbed, median and lateral body setae similar. Guard seta absent or not discernible on tarsus I.

Type species.—*Cheletophyes vitzthumi* Oudemans, 1914, Ent. Ber. Nederl. Ent. Ver. 4 (78): 101. Monotypic.

Cheletophyes eckerti, new species
(Fig. 32)

Palp claw with 3 basal teeth. Outer comb 8 or 9 short, bluntly rounded teeth; inner comb 6 short, thickened teeth. Axis of palp femur nearly straight; dorsal palp femoral seta (59) robust, about 1.4 times as long as longest dorsolateral seta of propodosoma. Rostrum projects forward to level of palp tarsi, apex a fairly pointed tip bearing a pair of very minute, outwardly projecting spines or sensilla. Stylophore approximately bell-shaped, heavily sclerotized, dorsal surface imbricate; protegmenal portion gently domed over stylet bases, anterior margins converging to fit into rostral gutter. Peritremes M-shaped, large in caliber, conspicuous; 11 very short bulbous chambers (segments) per side. Idiosoma pyriform in outline; propodosoma more massive than hysterosoma; the latter conical, bluntly pointed at rear. Dorsal plating comprises 9 median plates of unequal size and 9 pairs of very small platelets each bearing 1 seta; all plates heavily sclerotized, sharply outlined. Propodosomal plate quadrangular, widest and convex on rear margin; covers about 80 per cent of propodosomal tergum; embossed with a coarse reticulum of folds; numerous tiny pores, about 60, distributed near each side margin, others thinly scattered over midregion. Median hysterosomal plate small, ovoid, reticulate, covering only hindmost part of opisthosoma. Eyes with raised corneas. Integumental striae deeply etched, double-lined, tuberculate on dorsum, without tubercles on venter. Dorsal setae 15 pairs plus humerals, all one kind, relatively short (31–43), stout, rigid, spiniform; each with a dominant apical spikelet and few thick barbs distributed along midpart of shaft; 4th dorsolateral propodosomal seta on separate marginal platelet, close beside humeral seta; 1st to 4th dorsolateral hysterosomal setae on individual platelets, 5th and 6th dorsolaterals on suranal shield; dorsomedian setae (23–30) smaller than marginals, 2 pairs on propodosomal plate, 3 pairs (2½ pairs on holotype) on individual hysterosomal platelets. A pair of small ovoid depressed areas occur on hysterosoma, between 1st and 2d pairs of median setae. One pair of small, triangular floating sclerites occur in ventral integument, between mesal apices of coxae IV; these bear no seta. Cuticula of podomeres thick, tanned; all tarsi bear stout claws and multirayed empodia.

Length ratio: leg I/idiosoma = 0.8. Distribution of setae on legs I to IV: femora 2-2-2-1, genua 3-2-2-2, tibiae 6-4-4-4, tarsi 9-8-7-7. Tarsus I: solenidion wI (35) robust, tapered to pointed tip, on prominent nipple midway between proximal end and distal setae tc, without guard seta. Measurements (holotype only): length idiosoma 334, gnathosoma 148, leg I 226, tarsus I 94.

Holotype.—Female, from mite pouch of carpenter bee, *Xylocopa* (*Koptortosoma*) *aestuans* L., Punjab Agricultural University, Ludhiana, India (J. E. Eckert); deposited in UCD.

The genus *Cheletophyes* as currently constituted consists of but two species, *C. vitzthumi* Ouds. and *C. eckerti*, n. sp.

In *eckerti* the propodosomal plate is coarsely reticulate; the lateral hysterosomal setae are arranged four pairs on individual platelets and two pairs on the suranal shield; the dorsal setae on tibiae I–IV bear but a few basal barbs. In *vitzthumi* the propodosomal plate apparently is not reticulate; the lateral hysterosomal setae are arranged three pairs on individual platelets and three pairs on the suranal shield; the dorsal setae on tibiae I–IV are densely barbed from base to tip. Also, in *eckerti*, there is a pair of small sclerites ventrally, in the metasternal region, between coxae IV; *vitzthumi* also has a pair of small sclerites, but these are adjacent to coxae II.

Nodele Muma

Nodele Muma 1964, Florida Ent. 47 (4) :252.
Neocheletophyes Volgin, 1965, Akad. Nauk S.S.S.R., Trudy Zool. Inst. 35:296–297; Wafa and Soliman, 1968, Acarologia 10(2) :223–224.

Palp claw bears 1 prominent tooth on midsection. Palp tarsus with 2 sickles and 2 combs. Palp tibia lacks mesal flange overlying tibiotarsal joint. Protegmen subconic; comprises a substantial part of stylophore; its pointed apex presses into median gutter of rostrum. Dorsal body plates feebly sclerotized, boundaries indeterminate; humeral sulcus obsolete, not suture-like. Eyes present. Dorsal body setae long or very long, rodlike, profusely barbed; dorsolaterals and dorsomedians similar in form. With 2d pair of dorsomedian propodosomal setae set very close behind 1st dorsomedians. Leg I about as long as idiosoma. Length ratio: leg I/idiosoma = 0.9 to 1.1. Legs I to IV with paired claws and empodia; claws smooth, without basal apophyses.

Types species.—*Nodele calamondin* Muma; by original designation.

KEY TO FEMALES OF NODELE
(Based partly on Thewke and Enns, 1968)

1. Guard seta on tarsus I longer than solenidion wI; guard seta on separate papilla distant
  from papilla of wI 2
  Guard seta on tarsus I about half as long as wI; guard seta and wI juxtaposed on a single
  papilla 3
2. Guard seta with obvious barbs; from leaf litter, Egypt ....*simplex* Wafa & Soliman
  Guard seta smooth or nearly so; from leaf litter, U. S. A. ....*calamondin* Muma
3. With 1st pair of dorsomedian hysterosomal setae reaching only to bases of 3d pair of marginal setae ....*philippinensis* (Baker)
  With 1st pair of dorsomedian hysterosomal setae reaching to bases of 5th and 6th pairs of marginal setae ....*coccineae* Thewke & Enns

Nodele calamondin Muma

(Fig. 33)

Nodele calamondin Muma, 1964, Florida Ent. 47 (4) :252–253.

Outer comb ca. 30 teeth; inner comb ca. 30 teeth. Palp tibia carries claw on robust pedestal having no bladelike flange on its dorsomesal face. Protegmen subconic, merged; comprises a substantial part of stylophore in front of peritremes; its slender apical process(es) overlie stylets in rostral gutter. Tegmen and protegmen covered with longitudinal striae; striae thick and broken, pressed closely together. Rostrum well extended, spade-shaped; inferior adoral setae much shorter than superior adorals. Peritremes with 7 links per side, arranged in almost semi-circular arc, without abrupt changes in curvature. Dorsal plating very feebly sclerotized, boundaries indeterminate. Humeral sulcus obsolete. Dorsal setae of body and basal podomeres rodlike, abundantly barbed from bases to blunt tips; those of idiosoma relatively long (74–125); 10 pairs of dorsolaterals, 5 pairs of dorsomedians, all similar in form. Propodosoma with 2 pairs of dorsomedians; setae of posterior pair originate close behind (16) setae of anterior pair. Posterior seta of coxa I ultralong, flagelliform, long enough to extend across midventral line. Anterior seta on coxae III and IV rodlike, barbed; other coxals and intercoxals acicular, smooth. With 3 pairs of anal setae, dorsalmost (posterior) pair much longer than other 2 pairs.

Length ratio: leg I/idiosoma = 0.9, legs I/II = 1.2, legs I/IV = 0.9. Distribution of setae on legs I to IV; femora 2-2-2-1, genua 3-2-2-2, tibiae 6-5-4-4, tarsi 10-8-7-7. Tarsus I: solenidion wI (25) much shorter than smooth guard seta (59); wI arises on distal half of tarsus, remote from origin of guard seta. Measurements (n = 1): length idiosoma 410, gnathosoma 144, leg I 357, leg II 296, tarsus I 144.

Collection data.—From leaf litter of Monterey pine, U. C. campus, Davis, Calif. (D. W. Price).

Nodele philippinensis (Baker)

(Fig. 34)

Cheletophyes philippinensis Baker, 1949, Proc. U.S. Nat. Mus. 99(3238):288.
Nodele philippinensis (Baker), Muma, 1964, Florida Ent. 47(4):252.
Neocheletophyes philippinensis (Baker), Volgin, 1965, Akad. Nauk S.S.S.R., Trudy Zool. Inst. 35:296–297.

Outer comb ca. 23 teeth; inner comb ca. 26 teeth. Rodlike barbed setae on femur (2) and genu (1) of palp; most of ventral setae on palp segments and basis capituli smooth, flagelliform, some ultralong. An appreciable part of pointed rostrum projects from beneath a domed protegmen; surface of protegmen with a few tuberclate longitudinal ridges over midsection, with tubercles becoming more densely packed and lozenge-shaped at sides. Each peritreme comprises 8 or 9 links arranged in curved arc having no angular bends. Tegmen ornamented with closely set thickened striae broken into linear series of bacillus- or coccus-like swellings. Dorsal plating of idiosoma very feebly differentiated from marginal cuticle; according to regional variations in organization of striae, the body appears to have 2 large plates on which dorsal setae are implanted. Eyes present, wedged externally between tubercles of 2d and 3d marginal setae. Dorsal setae rodlike, blunt, strongly barbed, conspicuously long; 15 pairs plus humerals. Lateral setae of hysterosoma and median setae of propodosoma show marked tendency to cluster into groups of 2 setae set very close together. Median setae have same general form as laterals but with shorter and less acutely pointed barbs. With 1 seta on each coxa I and II and 4th pair ventral body setae exceptionally long (100–150), flagelliform. Paragenital setae 2 pairs, genitals 2 pairs, anals 3 pairs, posterior anals rodlike, barbed, other 2 pair smooth, flagelliform.

Length ratio: leg I/idiosoma = 1.1. Setae on podomeres: femora 2-2-2-1, genua 3-2-2-2, tibiae 6-5-4-4, tarsi 9-8-7-7. Solenidion wI (47) slightly less than half as long as tarsus (body); guard seta short (12), very slender, difficult to observe, close behind wI . Measurements (holotype): length idiosoma 403, gnathosoma 152, leg I 456, tarsus I 152, vertical seta 137, 1st median seta on propodosoma 114, humeral seta 220, hindmost dorsal seta 90.

This redescription is based on the holotype, USNM 1763.

The two most recently described species are:

Nodele coccineae Thewke and Enns, 1968, Acarologia 10(2):215–219. Known to occur in galleries of bark beetle, *Pseudopityophthorus minutissimus,* in scarlet oak.
Nodele simplex Wafa and Soliman, 1968, Acarologia 10(2):223–224. Taken from dried leaves, Giza, United Arab Republic.

Mexecheles De Leon

Mexecheles De Leon, 1962, Florida Ent. 45(3):132; Muma, 1964, Florida Ent. 47(4):248.
Acarocheyla Volgin, 1965, Akad. Nauk S.S.S.R., Trudy Zool. Inst. 35:293. New synonymy.

Palp claw with 7 to 12 teeth. Palp tarsus with 2 sickles and 2 combs. Dorsolateral seta on palp femur lanceolate to fanlike, monaxial or bifid. Protegmen well developed, subconic, apex merged with rostrum or front rim slightly elevated. Anterior half of tegmen decorated with lattice of tuberculate ridges; cells of lattice tend to be stretched crosswise; cells enclose small aggregates of microtubercles or patches of interrupted striae. Dorsolateral and dorsomedian body setae dissimilar; dorsolaterals lanceolate, paddle-shaped or straplike; dorsomedians fragmented or staghorn-like. Eyes present. With 2 or 3 pairs of paragenital setae. Leg I as long as or longer than idiosoma. Solenidion wI and guard seta conspicuously developed. Claws on tarsus I very small. Femur IV bears 2 setae.

Type species.—Mexecheles cunliffei De Leon; by original designation.

Volgin (1965) performed a useful service by clarifying the status of *Cheletophyes*. He erected *Acarocheyla* for *Cheletophyes hawaiiensis* Baker, 1949, but he did not deal with *Mexecheles cunliffei*, an allied species which De Leon (1962) set up as the type species of *Mexecheles*. The distinctions between *Mexecheles cunliffei* and *Acarocheyla hawaiiensis* become less obvious when five other allied species are studied. After examining the specimens described by Baker (1949), De Leon (1962), and Smiley and Moser (1970), we are more impressed with the similarities of the seven species than with the differences between the two type species, *cunliffei* and *hawaiiensis*. We therefore place *Acarocheyla* Volgin as a synonym of *Mexecheles* De Leon.

Mexecheles species may be recognized as fairly long-legged mites having very tiny claws on tarsus I, seven or eight pairs of fragmented dorsomedian setae, and paddle-shaped or straplike dorsolateral setae. The reticulation of the tegmen is also a useful identififying feature.

Mexecheles and *Cheletomorpha* appear to be closely allied genera.

KEY TO FEMALES OF MEXECHELES

1. Dorsolateral seta on palp femur bifid 2
   Dorsolateral seta on palp femur monaxial 5
2. Mesal paraterminal seta on tarsus I hypertrophied, extending far beyond tips of claws and
      empodial raylets 3
   Mesal paraterminal seta on tarsus I normal, not extending far beyond claws and raylets 4
3. All polygons in latticework on tegmen enclose numerous rounded microtubercles
                                                                    hawaiiensis (Baker)
   A few polygons in lattice with numerous rounded microtubercles; most others enclose
      patches of broken varicose striae ....*aztecorum* De Leon
4. With 1st dorsolateral hysterosomal setae long enough to reach base of 4th dorsolateral
      hysterosomal seta ....*marshalli* (Baker)
   With 1st dorsolateral hysterosomal seta not long enough to overreach base of 3d dorso-
      lateral hysterosomal seta ....*impolitus* Smiley and Moser
5. Claws on tarsi I to IV are smooth hooklets, having no basal outgrowths; no papillae on
      interscutal integument ....*cunliffei* De Leon
   Claws on tarsi I to IV bear apophyses or basal processes; numerous papillae on dorsal inter-
      scutal integument 6
6. Length of anteriormost dorsolateral body seta (i.e., vertical) considerably exceeds distance
      from its base to dorsal midline ....*virginiensis* (Baker)
   Length of anteriormost dorsolateral body seta obviously less than distance from its base
      to dorsal midline ....*panneus*, n. sp.

Mexecheles cunliffei De Leon

(Fig. 35)

Mexecheles cunliffei De Leon, 1962, Florida Ent. 45(3):132–133.

Palp claw with 8 or 9 teeth. Outer comb ca. 25 teeth, inner comb ca. 35 teeth. Dorsal seta of palp femur lanceolate, fluted and abundantly barbed; dorsolateral seta of this segment monaxial, fanlike, almost truncate across distal end. Cuticula of palp femur striate, without denticles or microtubercles. Rostrum almost pointed, with lateral papillae much reduced; exposed length about equal to midline length of protegmen. Protegmen subconic, slightly elevated across truncate front margin, ornamented with broken striae coursing longitudinally. Anterior half of tegmen displays coarse latticework of surface ridges; cells of lattice erratic in shape but oriented mostly with long axes in transverse plane; larger cells of lattice contain rows of broken striae, those within one enclosure tending to run in a different direction from those in adjoining cells; short bacillus-like striae cover tegmen above and behind peritremal grooves. Dorsal plates with faint, plain striae; interscutal membrane with simple striae, microtubercles and papillae absent. Dorsal body setae: 18 pairs plus humerals; dorsolaterals and dorsomedians dissimilar. Dorsolateral setae spatulate or paddle-shaped, relatively short (35–70), 4 pairs on propodosomal plate, 4 pairs on hysterosomal plate, 2 posteriormost pairs on opisthosomal integument. Dorsomedian setae fragmented, each comprising possibly 40 to 50 sclerotic particles strung on a very delicate framework of branched strands of matrix; 5 pairs on propodosoma, 3 pairs on hysterosoma. Anogenital area has 2 pairs of paragenital setae, 2 pairs of genitals, 3 pairs of anals, all simple, acicular; a pair of ultralong (78) flagelliform setae on venter of opisthosoma lie considerably in front of genital slit and, in this species, are not classed as paragenitals.

Length ratio: leg I/idiosoma = 1.2. Setae on legs I to IV: femora 2-2-2-2, genua 3-2-2-2, tibiae 6-4-4-4, tarsi 10-8-7-7. Tarsus I long and very slender, pedicel prominent, true claws diminutive; mesal paraterminal seta not projecting beyond tips of empodial raylets; upper face of tarsus rippled or undulant between nipple of *w*I and origins of addorsal setae; solenidion *w*I unusually long (113), overreaching bases of addorsals *tc*; guard seta (82) about four-fifths as long as *w*I, finely barbed. Claws on legs I to IV smooth hooklets, without apophyses.

M. cunliffei is redescribed from the holotype female, taken from *Ipomea murucoides*, Huajuapan de Leòn, Oaxaca, Mexico (D. De Leon); the specimen was loaned by the Museum of Comparative Zoology, Harvard University, by courtesy of Dr. H. W. Levi.

The type species happens to be one of the least specialized of the species so far referred to *Mexecheles*. The dorsolateral setae are conservative in length, the body striations are plain, the claws are without apophyses, and the mesal paraterminal seta of tarsus I is not hypertrophied. The rippled upper surface of tarsus I is an uncommon character.

Mexecheles hawaiiensis (Baker)

(Fig. 36)

Cheletophyes hawaiiensis Baker, 1949, Proc. U.S. Nat. Mus. 99(3238):289.
Mexecheles intermedius De Leon; Muma, 1964, Florida Ent. 47(4):248.
Mexecheles hawaiiensis (Baker); Muma, 1964, *ibid.*:248.
Acarocheyla hawaiiensis (Baker); Volgin, 1965, Akad. Nauk S.S.S.R., Trudy Zool. Inst. 35:293–294.

Palp claw bears 9 or 10 teeth. Outer comb ca. 20 teeth; inner comb ca. 35 teeth. Dorsal seta of palp femur rodlike, crudely pointed, coarsely barbed; dorsolateral seta bifid, well barbed, clearly inserted on femur. Protegmen subconic, its front apex barely evident; lattice of 8 to 10 polygonal cells covers dorsocentral area, each cell bounded by lightly elevated ridges of tuberculate striae, cells with greatest diameters in longitudinal axis of stylophore; relative lengths protegmen : tegmen = 10:13. Tegmen with lattice of about 16 irregularly shaped polygons, each cell bounded by dotted striae and enclosing numerous microtubercles; area of muscle attachment near base of stylophore devoid of granulations but covered with faint, broken striae. Peritremes with 5 or 6 cells per side. Two dorsal plates entirely cover upper surface of idiosoma, their surfaces abundantly stippled with microtubercles but with no obvious striations. Interscutal integument of dorsal and pleural areas covered with microtuberculate striae; microtubercles dispersed like those on plates; on edge view, cuticle shows microtubercles as very tiny erect rodlets which make entire dorsum rough. Nymphs of this species have numerous papillae on interscutal integument; adults have no obvious papillae. One pair of eyes, external to 3d marginal seta on propodosomal plate. Dorsal setae: 18 pairs plus humerals; 10 pairs dorsolaterals (27–90), straplike, ends abruptly rounded, with 1 or 2 barbed ridges on convex face; 8 pairs of dorsomedian setae fragmented or staghorn-like. Propodosomal plate bears 4 pairs of marginals (or dorsolaterals) and 5 pairs of fragmented medians; hysterosomal plate bears at least 4 pairs of marginals and 3 pairs of fragmented medians; posteriormost 2 pairs of marginals appear to originate on individual, trivial platelets. Setae on venter of body and on coxae I to IV flagelliform, smooth, exceptionally long, e.g., subcapitular 98, posterior coxal I 109, anterior paragenital 75 (minimal measurements). Paragenitals 3 pairs, of which anteriormost is longest; 2 pairs on genital covers; 3 pairs stout, stubby setae on semicircular anal swelling.

Length ratio: leg I/ idiosoma = 1.3. Setae on legs I to IV: femora 2-2-2-2, genua 2-2-2-2 (distal solenidion on genu I not readily identifiable), tibiae 6-5-4-4, tarsi 10-8-7-7. Tarsus I bears a long pedicel and diminutive true claws; azygos seta near base of pedicel reduced to a microseta; mesal paraterminal seta conspicuously thickened and lengthened (70); solenidion wI ca. 1.5 times as long as tarsus less pedicel; guard seta (55) ca. 0.5 times as long as wI; ventral seta v very short (20), smooth, set closer to addorsals than to nipple of wI. All paired claws smooth. Measurements (n = 7): length idiosoma 306 ± 16, gnathosoma 124 ± 5, leg I 397 ± 24, tarsus I (incl. pedicel and claws) 116 ± 4, vertical seta 81 ± 4.

Mexecheles aztecorum De Leon

(Fig. 37)

Mexecheles aztecorum De Leon, 1962, Florida Ent. 45 (3) :133–134.

Palp claw bears 8 to 11 teeth. Outer comb ca. 21 teeth; inner comb ca. 31 teeth. Dorsal seta on palp femur bifid, as in *hawaiiensis*. Middle surface area of protegmen with about 10 polygonal cells marked off by slightly elevated ridges; margins of protegmen densely stippled with coccoid microtubercles. Lattice of coarse polygonal cells also covers most of tegmen; a few cells adjacent to peritremes packed with microtubercles, but majority of cells enclose whorls of broken varicose striae. Interscutal integument of dorsum bears microtuberculate striae, not papillose. Two major plates cover most of dorsum; plating very lightly sclerotized, with whorls of microtuberculate striae resembling those of interscutal areas. Dorsolateral body setae: 10 pairs plus humerals, moderately long (39–90); 4 pairs on propodosomal plate, 4 pairs on hysterosomal plate, remainder on individual, trivial platelets. Dorsomedian setae aberrant, fragmented or staghorn-like; 4 pairs on propodosoma, 3 pairs on hysterosoma.

Length ratio: leg I/idiosoma = 1.0. Setae on legs I to IV: femora 2-2-2-2, genua 3-2-2-2, tibiae 6-5-4-4, tarsi 10-8-7-7. Tarsus with pedicel considerably extended, paired claws diminutive; mesal paraterminal seta conspicuously enlarged and long (55); azygos seta on base of pedicel reduced to a microseta; solenidion wI very long (78), overreaching bases of addorsals tc; guard seta finely barbed, conspicuous, about 0.7 times as long as wI; ventral seta very short (23), smooth, set closer to addorsals tc than to nipple of wI. All paired claws smooth. Measurements (n = 1): length idiosoma 410, gnathosoma 129, leg I 403, tarsus I 125, vertical seta 90.

Collection data.—Illustrated specimen, a paratype, collected on oak, *Quercus conzattii*, Oaxaca, Oaxaca, Mexico (Donald De Leon).

M. aztecorum is a close match to *hawaiiensis,* and the differences between the two are subtle. In *aztecorum* the majority of the polygonal cells on the tegmen enclose whorls of broken striae; in *hawaiiensis* the closed polygonal cells are uniformly stippled with microtubercles. The distal solenidion on tibia ɪ in *aztecorum* is short, barely reaching to the line of articulation between tibia and tarsus; in *hawaiiensis* this solenidion extends about half its length beyond the line of articulation. The available paratype specimen of *aztecorum* has four pairs of fragmented median setae on the propodosomal plate; the common number of fragmented median setae on the propodosoma of *hawaiiensis* is five pairs. Although Baker (1949) noticed only four pairs of these setae on the holotype female of *hawaiiensis,* we are able to find an additional pair on this specimen probably because some of the fragmented particles were broken off when the specimen was subsequently re-mounted.

Mexecheles panneus, new species

(Fig. 38)

Palp claw bears 10 to 12 teeth, 4 or 5 distal teeth in an offset row not quite aligned with 6 or 7 basal teeth. Outer comb with ca. 30 fairly long, coarse teeth; inner comb with 30 short, fine teeth. Palp tarsus an erect cylindrical appendage, about twice as long as it is wide. Palp tibia with unusually prominent straight-edged flange overlying tibiotarsal articulation. Outer face of palp femur sharply elbowed, its lateral margin turning forward in a 90° bend. Dorsal and lateral setae on palp femur fusiform, barbs in close-set, compact rows. Protegmen subconic, merged, comprising a substantial part of stylophore anterior to peritremes; it narrows abruptly in front to form an apical extension which appears to cover, or lie within, a rostral gutter; upper cuticle with 4 or 5 pairs of longitudinal tuberculate ridges. Dorsal cuticle of tegmen coarsely latticed with tuberculate ridges; about 8 completely closed meshes, several others imperfectly circumscribed, all containing fine, mostly transverse, varicose striae and a very few scattered tubercles. Peritremes shaped like inverted horseshoe, 4 slender links in each transverse arm, 3 or 4 links in each descending arm. Dorsal plating lightly tanned, very faintly striated, without tubercles or other noteworthy sculpturing. Hysterosomal plate covers most of tergal area of this tagma, about 0.8 times as broad as propodosomal plate. Interscutal integument papillose, with smooth, simple striae. Dorsolateral setae (35–59) narrow spatulate, considerably longer than greatest width (anterior propodosomals ca. 4 to 1), each with 1 to 3 barbed ribs; 4 pairs on propodosomal plate, 5 pairs on hysterosomal plate, 1 pair on integument near tip of opisthosoma; 3d dorsolateral propodosomals set just above eyes; 1st and 2d dorsolateral hysterosomals close together, 3d and 4th pairs far apart. Dorsomedian setae aberrant, each comprising a short, weak stalk and an aggregate of small angular sclerites, 15 to 30 perhaps, with a few sclerites spaced almost equidistant on each of several intertwined strands of hyaline matrix; 5 pairs on propodosomal plate, 4 pairs on hysterosomal plate. Posterior seta of coxa ɪ ultralong (105), flagelliform, long enough to cross midventral line. Anterior seta on coxae ɪɪɪ and ɪV fusiform, abundantly barbed; all other coxal setae acicular, smooth, drawn to exquisitely fine points. With 2d pair intercoxal setae also unusually long (75). With 3 pairs setae in paragenital zone, anterior-most pair excessively long (60); 3 pairs anal setae stubby, equispaced in arc across rear of anal papilla.

Leg ɪ relatively long, especially in relation to leg ɪɪ. Length ratios: leg ɪ/idiosoma = 1.0, legs ɪ/ɪɪ = 1.5. Claws on tarsi ɪɪ to ɪV with basal apophyses. Distribution of setae on legs ɪ to ɪV: femora 2-2-2-2, genua 3-2-2-2, tibiae 6-5-4-4, tarsi 10-8-7-7. Tarsus ɪ: solenidion *w*ɪ ultralong (78), reaching to claws; guard seta barbed, bushy, about 0.5 times as long as *w*ɪ; addorsals *tc* roughened or minutely annulate. Measurements (n = 2): length idiosoma 391, gnathosoma 140, leg ɪ 384, leg ɪɪ 247, tarsus ɪ 113.

Holotype.—Female, from juniper foliage, Lehman Creek, 8 miles W of Baker, Nevada (D. W. Price); deposited in UCD.

Mexecheles virginiensis (Baker)

(Fig. 39)

Cheyletia virginiensis Baker, 1949, Proc. U.S. Nat. Mus. 99(3238):299–300.
Paracheyletia virginiensis (Baker); Volgin, 1955, Akad. Nauk S.S.S.R., Zool. Inst., Opredel.
  p. Faune S.S.S.R. no. 59:169.
Mexecheles virginiensis (Baker); De Leon, 1962, Florida Ent. 45(3):132.
Acarocheyla virginiensis (Baker); Smiley and Moser, 1970, Proc. Ent. Soc. Wash. 72(2):229–
  236.

M. virginiensis resembles *panneus* in structural detail; their differences are essentially quantitative. The characters of *virginiensis* are more emphatically developed.

Dorsolateral setae narrow spatulate, relatively long (74–137), anteriormost propodosomals much longer than wide (proportions ca. 6 to 1); 2 to 4 heavily barbed ribs, membrane between ribs delicately net-veined. With 1st and 2d dorsolateral propodosomals close together, 3d and 4th pairs far apart. Dorsomedian setae 9 pairs, fragmented, patterned like those of *panneus* except that small angular sclerites are more numerous, possibly 50 or more sclerites per seta; 5 pairs on propodosoma, 4 pairs on hysterosoma. Interscutal integument papillose and striated; papillae tend to have very sharply pointed apices, in profile like rows of saw teeth.

Length ratio: leg I/idiosoma = 1.1, legs I/II = 1.3. Tarsus I: apophyses on claws of this leg quite prominent (claws on 1st leg of *panneus* not provided with distinct, readily discernible outgrowths); solenidion wI (100) almost long enough to reach claws. Measurements (n = 4): length idiosoma 516, gnathosoma 182, leg I 540, leg II 408, tarsus I 173.

Collection data.—From pine bark, associated with *Ips*, Incline Beach, Lake Tahoe, Washoe Co., Nevada; from tree bark, Beaver Mountain, Logan, Utah.

Mexecheles marshalli (Baker)

(Fig. 40c–e)

Cheletophytes marshalli Baker, 1949, Proc. U.S. Nat. Mus. 99(3238):290.
Acarocheyla marshalli (Baker); Volgin, 1965, Akad. Nauk S.S.S.R., Trudy Zool. Inst. 35:294–
  295.
Mexecheles marshalli (Baker). New combination.

Palp claw with 8 to 10 blunt teeth distributed on basal half, distal tooth perceptibly larger than others in this series. Outer comb ca. 26 teeth; inner comb ca. 35 teeth. Dorsal seta of palp femur has flattened blade, truncate diagonally so that it tapers to a point; dorsolateral seta on this segment also flattened and bifid. Protegmen subconic or domed, anteriormost limit defined by a faint, convex membranous fold. Relative lengths of protegmen: tegmen = 5:7. Upper surface of entire stylophore bears longitudinal rows of tuberculate striae, each interspersed between 2 or 3 rows of very delicate broken striae having no emphatic tubercles; pattern on protegmen repeated on tegmen. Peritremes about as illustrated, configuration of links not clearly discernible on holotype. Two plates cover entire dorsum of idiosoma; plating stippled with microtubercles aligned in ill-defined transverse rows. Interscutal integument of back and sidewalls with microtuberculate striations and a moderate number of rounded papillae. Dorsal setae: probably 17 pairs plus humerals; 10 pairs very elongate (to 153), straplike. A portion of opisthosoma is torn away on holotype; it is supposed that the missing part bears 2 pairs of marginals behind opisthosomal plate. Propodosomal plate bears 4 pairs of straplike marginals plus 4 pairs of fragmented medians fashioned as in *hawaiiensis*; hysterosomal plate bears at least 4 pairs of straplike marginals plus 3 pairs of fragmented medians. Ventral body setae and those on coxae I to IV very long, flagelliform, e.g., posterior seta of coxa I 100, posterior seta of coxa II 75, anterior paragenital at least 75. With 3 pairs of paragenitals, 2d and 3d pairs relatively short, with frayed ends.

Length ratio: leg I/idiosoma = 1.3. Setae on legs I to IV not determinable. Tarsus I with very long pedicel and diminutive claws; mesal paraterminal seta not unusually developed; solenidion wI (105) about as long as body of tarsus less pedicel; guard seta (101) in duplex position on nipple, profusely barbed, about as long as wI. Claws on legs I to IV smooth, without apophyses. Measurements (n = 1), partly from Baker's original description: length idiosoma 300, gnathosoma 133, leg I 406, tarsus I 148, vertical seta 153, 1st marginal seta on hysterosomal plate 140.

Collection data.—Type specimen, USNM 1765, taken at Imboden, Arkansas; the only known specimen. This remounted specimen has some parts missing and is otherwise in poor condition.

M. marshalli shares several features with *virginiensis* and *panneus:* the interscutal integment of the back and sides bears papillae as well as tuberculate striae; the mesal paraterminal seta on tarsus I is not hypertrophied. It also has several characters in common with *hawaiiensis* and *aztecorum:* the claws on legs I–IV are smooth hooklets and the dorsolateral seta on the palp femur is bifid.

The dorsal and lateral setae on the palp femur are broader, the lateral seta of this segment is much longer, and the ornamentation of the stylophore differs from that described for all the other species mentioned above.

Mexecheles impolitus (Smiley and Moser)
(Fig. 40*a, b*)

Acarocheyla impolitus Smiley and Moser, 1970, Proc. Ent. Soc. Wash. 72(2):229–236.
Mexecheles impolitus (Smiley and Moser). New combination.

This species is an ally of *marshalli* and is possibly identical with it in chaetotaxy and ornamentation of the gnathosoma. It has, however, a different combination of body and leg characters.

Interscutal integument with microtuberculate striae and rounded papillae. Dorsal setae: 18 pairs plus humerals. Lateral setae (35–98) straplike but shorter than in *marshalli*. With 5 pairs of fragmented median setae on propodosoma, 3 pairs on hysterosoma. Length ratio: leg I/idiosoma = 1.2. Setae on legs I to IV: femora 2-2-2-2, genua 3-2-2-2, tibiae 6-4-4-4, tarsi 10-8-7-7. Tarsus I wavy or undulant in profile on dorsal face between wI and addorsals tc; wI appreciably curved near distal third, shorter than body of tarsus less pedicel; guard seta (74) almost equal to wI in length, coarsely barbed; mesal paraterminal seta acicular, barely reaching to tips of claws; claws of leg I minute. Claws of legs I to IV smooth, without apophyses. Measurements (n = 4): idiosoma 367, gnathosoma 134, leg I 425, tarsus I 154, vertical seta 91, 1st marginal seta on hysterosomal plate 78.

Collection data.—Holotype female taken from boring dust of *Dendroctonus frontalis* in loblolly pine, Elizabeth, Louisiana (J. Moser).

Grallacheles De Leon

Grallacheles De Leon, 1962, Florida Ent. 45(3):135; Muma, 1964, Florida Ent. 47(4):248.

Palp claw with 6 to 9 teeth. Palp tarsus bears 2 sickles and 2 combs. Protegmen elevated, convex in front, about half as long as tegmen. Appendages carry numerous foliate setae, some very broad and fanlike, others spatulate or lanceolate. Eyes present. Two major plates on dorsum. Lateral and median dorsal body setae predominantly straplike. Anteriomost lateral setae on propodosomal plate flabellate, 2d pair (preoculars) transitional, 3d and 4th pairs straplike. Posterior and setae markedly spatulate. Legs shorter than body, with noticeably short tibiae. Tarsi I to IV constrict to slender tubes, with paired claws and multirayed empodia. Tarsus I bears a very short, inconspicuous solenidion wI and a much enlarged spatulate guard seta.

Type species.—*Grallacheles bakeri* De Leon, 1962; type by original designation.

Grallacheles bakeri De Leon

(Fig. 41)

Grallacheles bakeri De Leon, 1962, Florida Ent. 45(3):135–137; Muma, 1964, Florida Ent. 47(4):248–250.

Paracheyletia woolfordi Cunliffe; Volgin, 1969, Akad. Nauk S.S.S.R., Zool. Inst., Opredel. p. Faune S.S.S.R. no. 101:275.

Palp claw bears 6 conical teeth (6 to 9 according to De Leon) equally spaced on basal half. Palp tarsus with 2 sickles, 2 combs. Outer comb ca. 22 teeth; inner comb ca. 30 teeth. Protegmen elevated, front margin slightly convex, upper surface longitudinally striated. Tegmen about twice as long as protegmen, with plain longitudinal striae. Peritremes inconspicuous, possibly 7 links per side; first 3 links of right and left peritremes form almost straight line across stylophore before curving abruptly rearward. Palps and ambulatory legs overlaid with foliate setae, a few of these almost flabellate, majority fanlike to spatulate. With 2 smooth plates covering most of dorsum; plating not obviously ornamented. Eyes present, external to bases of 2d and 3d marginal setae. Dorsal body setae: 15 pairs inserted on tergal plating plus 1 pair of humerals on independent marginal platelets; 4 pairs of laterals plus 3 pairs of medians on hysterosomal plate. Median setae and most of laterals are fluted bars with upturned margins, i.e., straplike; shorter ones grade into spatulate or fanlike shapes on opisthosoma. Anteriormost laterals (verticals) expanded into very prominent flabellae (62); 2d laterals transitional (74); 3d (121) and 4th (160) pairs straplike; 4th pair of laterals on propodosoma are longest dorsals. Genital area displays 2 pairs of paragenitals, 2 pairs of genitals, 3 pairs of anals; posterior anals markedly spatulate.

Length ratio: leg I/idiosoma = 0.7. Setae on legs I to IV: femora 2-2-2-1, genua 3-2-2-2, tibiae 6-4-4-4, tarsi 9-8-7-7. Tibiae I to IV diminutive, each not noticeably longer than its greatest diameter; each tibia bears 2 ventral setae, 1 spatulate, the other flagelliform. Each tarsus I–IV bears paired claws, rayed empodia, and a bilaterally cirrate ventral seta which partly ensheaths attenuated part of segment. Tarsus I with *w*I very short, inconspicuous; guard seta spatulate, almost as long as body of tarsus. Measurements. (n = 1): length idiosoma 350, gnathosoma 121, leg I 235, tarsus I 113.

Collection data.—Illustrated specimen from bark of citrus trees, Parrish, Florida (M. H. Muma). We have also examined holotype slide of *Paracheyletia woolfordi* Cunliffe, USNM 2882, from *Cycas revoluta*, Japan, intercepted at Hawaii (H. A. Woolford).

Cheletomorpha Oudemans

Cheletomorpha Oudemans, 1940*b*, Ent. Ber. Nederl. Ent. Ver. 1(18):162.

Palp tarsus bears 2 sickles and 2 combs. Protegmen elevated, forming elongate, semitransparent hood covering about two-thirds of spadelike rostrum. With 2 median shields covering most of tergum. Eyes present. Dorsolateral body setae and numerous appendicular setae long, stout, rodlike, blunt-tipped, densely barbed. Dorsomedian setae aberrant, few in number, inconspicuous; each comprises several moniliform strands in type species. Tarsus I with multirayed empodium at tip of a very long pedicel; true claws usually but not always lacking, when present minute; tarsi II to IV with paired claws and empodia. Addorsal setae *tc* and mesal paraterminal seta on tarsus I abnormally long; guard seta much longer than solenidion *w*I, both arising on same nipple.

Type species.—*Acarus lepidopterorum* Shaw, 1794. Monotypic.

Cheletomorpha lepidopterorum (Shaw)

(Fig. 42)

Acarus lepidopterorum Shaw, 1794, Nat. Misc. vol. 6, pl. 187.

Cheyletus venustissimus Koch, 1839, Deutschlands Crustaceen, Myriapoden und Arachniden, fasc. 23.

Cheyletus seminivorus Packard, 1878, A Guide to the Study of Insects, p. 665.

Cheyletus longipes Mégnin, 1878, Jour. Anat. et Physiol., 14:416–441.

Cheletomorpha venustissima (Koch); Oudemans, 1904*b*, Ent. Ber. Nederl. Ent. Ver. 1(18):162; Oudemans, 1906, Mém. Soc. Zool. France 19:144–153; Rohdendorf, 1940, Moscow Univ. Uchenye zapiski, Wissensch. Ber. 42:92–94.

Cheyletus rufus Hardy, in André, 1933, Ann. des Epiphyt. 19(6):352.

Cheletomorpha lepidopterorum (Shaw); Oudemans, 1937, Kritisch Historisch Overzicht der Acarologie. Derde Gedeelte, Band C: 1115–1117.

Cheletophyes tatami Hara, 1955; Volgin, 1969, Akad. Nauk S.S.S.R., Zool. Inst., Opredel. p. Faune S.S.S.R. no. 101:190.

Cheletophyes knowltoni Beer and Dailey, 1956; Volgin, 1969, *ibid.*: 190.

Gnathosoma projects well in front of body; basis capituli fairly long, supported on conical extension of propodosoma. Palp claw slender, obtusely bent near base, tip almost straight and spinelike, about one-third of total palp length; mesal flange bears 1 rounded tooth on basal end. Outer comb ca. 22 teeth plus strong terminal spine; inner comb ca. 30 teeth. Palp femur bulges outward, forming 90° elbow on lateral profile. Protegmen elevated, forming elongate hood over proximal two-thirds of rostrum, its front margin concave or with slightly extended marginal lobes. Entire stylophore and dorsal body plates similarly ornamented with granulate striae; granules minute, irregularly spaced in each row. Tegmen has short suture or cleft arising between mesal origins of peritremes. Peritremes approximately horseshoe-shaped, 6 links per side. Longitudinal apodeme on midventral line of basis capituli bifurcates as it passes forward between subcapitular setae. Rostrum distended ventrally to accommodate unusually capacious pharyngeal pump; tip rounded, somewhat spade-shaped; inferior adoral setae (50) much longer than superior adorals (20), inferior pair originates considerably behind superior pair. With 2 extensive shields almost covering entire dorsum. One pair of eyes with protruding corneas, encircled by 5 or 6 concentric striae. Dorsal body setae: 15 pairs (probably) plus humerals. Dorsolateral setae stout, rodlike, blunt-tipped, densely barbed, all moderately long (43–125); 4 pairs on propodosomal plate, 3 of which are tightly clustered near anterolateral angles of this plate; 5 pairs on hysterosomal plate, 1 pair on opisthosomal membrane. Dorsomedian setae aberrant, 2 + 3 pairs on front and rear plates, respectively; each seta a bouquet of several moniliform strands branching from a minute stalk. Only 1 of 4 pairs of ventral opisthosomal setae occur in paragenital position; 2 pairs of genitals; 3 pairs of anals.

Forelegs especially long and slender. Length ratios: leg I/idiosoma = 1.4; leg I/leg II = 1.9. Setae on legs I to IV: femora 2-2-2-1, genua 3-2-2-2, tibia 6-5-4-4, tarsi 10-8-7-7. Tarsus I: addorsals *tc* ultralong, with ridgelike annulations; pretarsus comprises extremely small multirayed empodium, with true claws absent in most specimens; pedicel attenuate; mesal paraterminal seta exceptionally long (95), at least 5 times as long as lateral paraterminal; guard seta a profusely barbed bristle, about as long as entire tarsus; guard seta arises close beside much shorter solenidion *w*I (50). Measurements (n = 10, local populations): length idiosoma 485±69, gnathosoma 148±11, leg I 655±79, tarsus I 175±13.

Collection data.—California: from cattle feed bins, dairy barn floors, Davis (D. W. Price); soil mulch, Sonoma (W. H. Lange); bark of ponderosa pine, Yosemite Valley (D. W. Price); hindwing of *Proxenus* moth (Noctuidae), Yolo (H. Michalk).

Good series of specimens in the above-mentioned collections contain several males which can be recognized as the opposite sex of the females examined. In the male an external lamella arches across the basis capituli, between opposite palp trochanters. This lamella is notched in the midline to form a pair of projecting lobes. In this respect, as well as in most details of female anatomy, the species described here is a good fit with Oudeman's (1906) account of *Cheletomorpha venustissima,* the species which he later (1937) called *lepidopterorum.* Oudemans noticed the peculiarities of the tarsus and the absence of claws on this leg, but he apparently did not recognize the strange dorsomedian setae.

The four setae illustrated in figure 42*e* show that the second moult, deutonymph to adult in the female line, may result in an incomplete transformation of the dorsomedian setae. The orthodox form of the dorsomedians in the deutonymph (bottom, left) normally transforms into the strandlike structure most characteristic of adult females (uppermost figure, and lower right). The imperfectly transformed seta (bottom, middle) is an oddity.

The mites which Beer and Dailey (1956) called *Cheletophyes knowltoni* probably represent long-spined variants of *Cheletomorpha lepidopterorum*. Four females in the paratype series of *knowltoni* are not qualitatively distinguishable from those which others have identified as *lepidopterorum*. Those which we have illustrated as *lepidopterorum* from California and those which Beer and Dailey illustrated from

TABLE 1

RANGES OF OBSERVED MEASUREMENTS (IN MICRONS) WITHIN THREE GROUPS
OF *Cheletomorpha lepidopterorum* (SHAW)

Group	No. of females observed	Length				
		Seta *ve*	*ve* to *ve*	Solenidion *w*I	Guard seta	Tarsus I over-all
California......	4	70–98	98–117	47–55	164–183	179–183
Kansas.........	4	109–121	109–129	82–98	176–183	183–191
World-wide.....	10	74–144	90–129	35–105	164–199	144–203

Kansas have different measurements for some of their parts. Several of the setae of the Kansas specimens are appreciably longer than those of our California specimens. But the measurement data obtained from small samples of specimens, as available, do not give an adequate basis for separating two species.

Table 1 summarizes the range of several lengths and distances observed for three assortments of specimens. The "world-wide" category contains individuals from each of ten different localities on three continents. The measurements made for three of the characters do not differentiate between locality samplings, i.e., length of tarsus I, length of guard seta, and distance between vertical setae *ve-ve*. On the other hand, the lengths of the vertical setae *ve* and the solenidia *w*I of specimens collected in Kansas differ significantly from those of specimens collected in California. However, both sets of measurements lie within the size range of the "world-wide" category.

These samples of *lepidopterorum* show another kind of variation. Some of the specimens show small paired claws on tarsus I; others do not. Three of eighteen specimens examined have claws; two females in the *knowltoni* type series have them and two do not. The presence or absence of claws in a species where claws are small enough to be vestigial, perhaps functionless, may not be as reliable for specific diagnoses as hitherto supposed.

The publication of the original descriptions of two other species is given below:

Cheletomorpha orientalis Oudemans, 1928, Ent. Ber. Nederl. Ent. Ver. 7(162):343. On leaves of orchid, *Phalaenopsis* sp., Java.

Cheletomorpha bakeri Lawrence, 1954, Ann. Natal Mus. 13(1):67. Taken from nest of a weaver bird which was used by *Galago senegalensis*, Karamoja, Uganda.

The dorsomedian setae of females of this species resemble the dorsolaterals; there is one pair of dorsomedians on the propodosomal plate and there are two pairs on the hysterosomal plate. The dorsomedian setae of *lepidopterorum* females are very much modified, and there are two and three pairs on these plates, respectively. *C. bakeri* has two pairs of spatulate anal setae; none of the anals of *lepidopterorum* are spatulate.

Cheletomorpha bakeri was placed in a separate genus, *Acheletomorpha*, by Volgin (1969). However, *C. lepidopterorum* and *C. bakeri* show so many features in common that it may be more prudent to regard their few differences as having specific rather than generic significane.

Chiapacheylus De Leon

Chiapacheylus De Leon, 1962, Florida Ent. 45(3):135.

Palp claw toothless. Palp tarsus with 2 combs, 2 sickles. Setae of appendages and upper body foliaceous; those of tergum flabellate or cycloid, strongly ribbed, with sharp marginal spikelets. Two eyes. Dorsum entirely covered with 2 plates, each having several deep, sinuous grooves. Propodosomal plate bears 6 pairs of lateral setae. Dorsolateral and dorsomedian setae alike in form, the latter almost as numerous as the former. Legs short in relation to body length. Tarsus I bears a multirayed empodium on a very stubby pedicel, with no true claws. Legs II to IV have paired claws and empodia.

Type species.—Chiapacheylus edentata De Leon; by original designation.

Chiapacheylus edentatus De Leon
(Fig. 43)

Chiapacheylus edentatus De Leon, 1962, Florida Ent. 45(3):135.

Dorsal palptibial seta broad, leaflike; partly enshrouds palptarsal sensilla. No teeth on palp claw. Outer comb ca. 12 teeth; inner comb ca. 16 teeth. Protegmen moderately elevated, about 0.8 times as long as tegmen; both areas of stylophore with plain linear striae. Eyes present. Dorsal body setae flabellate, cupped and ribbed, ribs terminating in sharp marginal spikelets; 23 pairs plus humerals; laterals and medians similar in form. Propodosomal plate bears 6 pairs of laterals and 5 pairs of medians; hysterosomal plate bears 5 pairs of laterals and 6 pairs of medians; last pair of laterals appear to originate on trivial platelets behind principal plate. Setal sockets on main plates insert on small elevated ovoid areas which seem to be optically different from surrounding sclerotized cuticula. Anogenital region with 2 pairs of very fine paragenital setae, 2 pairs of short genitals, 3 pairs of anals; 1st and 2d anals bifid, 3d pair fanlike.

Length ratio: leg I/idiosoma = 0.5. Tarsus I has a very broad guard seta (31 long) set considerably behind a short solenidion wI (ca. 8); pedicel stubby, no true claws, empodium erect, multirayed. Measurements (n = 3); length idiosoma 201, gnathosoma 58, leg I 100, tarsus I 35.

*Collection data.—*Holotype specimen collected on *Jaquinia pungens*, Tuxtla Gutierrez, Chiapas, Mexico (Donald De Leon).

The material examined by us included one female, from *Xylosoma elliptica*, Tuxtla Gutierrez, Chiapas, Mexico, and two females from *Citharexylum fructicosum*, Juana Diaz, Puerto Rico, all collected by Dr. De Leon. None of these specimens were designated as paratypes.

Two Egyptian species were described in *Chiapacheylus* by Zaher and Soliman (1967). These species, *Chiapacheylus desertorum* and *C. macrocorneus*, should not be included in *Chiapacheylus* De Leon as redefined here. We think that these Egyptian species may belong in *Hemicheyletia* Volgin, 1969, but without firsthand knowledge of the type specimens, we are unable to deal constructively with this problem.

Cheletogenes Oudemans

Cheletogenes Oudemans, 1905, Ent. Ber. Nederl. Ent. Ver. 1(21):208.

Palp claw with numerous minute teeth distributed along most of inner margin. Palp tarsus with 2 sickles, 2 combs. Protegmen elevated, bulbous. Dorsal and pleural cuticula bears numerous equispaced papillae. Dorsal plating also papillose. Propodosomal plate fairly well defined, covering most of propodosoma. Hysterosomal plating ill defined; appears to be restricted to midportion of metapodosoma, supporting 1st and 2d pairs of dorsomedian setae. Dorsal setae cycloid or rounded fans; dorsomedians and dorsolaterals alike. Some ventral body setae ultralong, flagelliform. Anal setae acicular, smooth or nearly so, none fan-shaped. Legs much shorter than body. Tarsus I stubby, truncate on distal end, with no pedicel or pretarsal structures; bears only 5 setiform sensilla: 2 long subequal setae in terminal position, 2 subapical microsetae, 1 small solenidion wI set on elevated nipple (fig. 44d), without guard seta. Tarsi II to IV equipped with pedicel, claws, and empodium.

Type species.—Cheyletus ornatus Canestrini and Fanzago, 1876. Monotypic.

Until recently this genus comprised an assemblage of distant relatives grouped together somewhat artificially according to the structure of tarsus I. Volgin (1969) believes that the species subsequently referred to *Cheletogenes* by their authors were not congeneric with *ornatus*. He created *Prosocheyla* for five of these species and left *Cheletogenes* with only its type species.

Cheletogenes ornatus (Can. and Fanz.)

(Fig. 44)

Cheyletus ornatus Canestrini and Fanzago, 1876, Atti Soc. Veneto-Trentina Sci. Nat. 5:106.
Cheyletus saccardianus Berlese, 1886; Oudemans, 1906, Mém. Soc. Zool. France 19:153.
Cheyletia ornata (Canestrini and Fanzago); Oudemans, 1904a, Ent. Ber. Nederl. Ent. Ver. 1(17):154.
Cheyletus cocciphilus Banks, 1914; Baker, 1949, Proc. U.S. Nat. Mus. 99 (3238):305.
Cheletogenes ornatus (Canestrini and Fanzago); Oudemans, 1905, Ent. Ber. Nederl. Ent. Ver. 1(21):208; Oudemans, 1906, Mém. Soc. Zool. France 19:153–158; Baker, 1949, Proc. U.S. Nat. Mus. 99(3238):305–306; Volgin, 1955, Akad. Nauk S.S.S.R., Zool. Inst., Opredel. p. Faune S.S.S.R. no. 59:174

A relatively small rotund species having a papillose body integument and short palplike forelegs. Palp claw with ca. 15 teeth, distributed along most of inner margin. Outer comb with ca. 20 teeth; inner comb with ca. 25 teeth, similarly fashioned. Inner sickle appears to be a flattened hook. Dorsal palpfemoral seta asymmetrically shaped, its mesal edge drawn into a pointed process which loosely enwraps palp tarsus. Cancellate areas of muscle attachment on dorsolateral face of palp femur bordered in front with a row of superficial microtubercles. Protegmen bulbous or domed, covering about three-fourths of rostrum; tegmen and protegmen ornamented with a whorl of irregularly thickened, broken striae. Peritremes with 5 to 7 links; posteriormost link curves sharply toward midline. Entire dorsal and pleural cuticula of idiosoma provided with numerous papillae. With 2 dorsal plates; propodosomal plate clearly covers area bounded by 4 pairs of marginal setae; hysterosomal plating ill defined, probably restricted to metapodosoma, supports only first 2 pairs of median setae. Dorsal body setae: 15 pairs (12–39) plus humerals; no structural difference between laterals and medians; each seta comprises a short pedicel and a scalelike or cycloid blade having about 10 coarsely barbed ribs; 4 laterals plus 3 median pairs on propodosomal plate, 6 laterals plus 2 median pairs on hysterosoma. With 1st, 2d, and 4th pairs of ventral body setae ultralong, flagelliform; 3d pair relatively short; 4th pair ($>$ 100) far over-reaches posterior margin of opisthosoma. Anogenital setae simple, 2 + 2 + 3 pairs paragenitals, genitals, anals, respectively.

genua 3-2-2-2, tibiae 5-4-4-4, tarsi 5-8-7-7. With 2 setae on femur IV and 5 setae on tarsus I, which

Length ratio: leg I/idiosoma = 0.5, leg I/leg II = 1.0. Setae on legs I to IV: femora 2-2-2-2, is unusual. Tarsus I short, truncate on distal end, pretarsus and pedicel lacking; 2 long addorsal setae *tc* anchored on its blunt end; 1 pair ventral microsetae in subapical position; solenidion *w*I very short (10), set on a prominent nipple; guard hair absent. Measurements (n = 10): length idiosoma 265 ± 31, gnathosoma 79 ± 4, leg I 130 ± 9, leg II 130 ± 11, tarsus I 25 ± 2.

Collection data.—Almond twigs, Altadena, Calif. (E. I. Schlinger); pansy foliage, Davis, Calif. (E. I. S.); oak twigs, Green Valley, Solano Co., Calif. (S. F. Bailey); lichens, Davis, Calif. (A. Beck); lichens, Pismo Beach, Calif. (F. C. Raney); bark of peach trees, Fowler, Calif. (F. M. Summers); bark of prune trees, Napa, Calif. (J. Skelsey); juniper foliage, Holbrook, Nevada (F. M. S.); sagebrush, Logan, Utah (G. F. Knowlton, E. H. Kardos); on *Vallesia*, Darwin Research Sta., Galapagos Islands (I. Wiggins); on *Castela* sp. and *Tournefortia* sp., Santa Cruz Is., Galapagos Islands (R. O. Schuster).

The structure of tarsus I and its simplified chaetotaxy and the papillose dorsal integument are distinctive characters.

Prosocheyla Volgin

Prosocheyla Volgin, 1969, Akad. Nauk S.S.S.R., Zool. Inst., Opredel. p. Faune S.S.S.R. no. 101:288–289.

Palp claw bears 3 to 13 teeth. Palp tarsus with 2 combs and 2 sickles. Protegmen elevated, with prominent front fold, or subconic and merged with rostrum. Eyes present. Dorsal body plating variable in area or definition; may be obsolete on both parts of idiosoma, or hysterosomal plating may be reduced to 1 or 2 small median plates. Dorsal setae short, fanlike, of medium width. Dorsomedian setae present, orthodox or aberrant, the latter lubulate (fig. 47) or fragmented (fig. 46*b*). Legs shorter than idiosoma, II to IV with claws and empodia. Tarsus I truncate at tip, with 4 to 6 setae arising on or near blunt end; solenidion *w*I at least half as long as body of tarsus; guard seta minute or absent.

Type species.—*Cheletogenes oaklandia* Baker; by original designation.

The identifying feature of the species presently assigned to *Prosocheyla* is the arrangement of the distal setae on tarsus I. This podomere has no tapered pedicel or pretarsal appendages; it ends bluntly and bears at least four conspicuous setiform sensilla and, in most species, one or two additional small subterminal setae.

Even with *Cheletogenes ornatus* removed from the group of species now accommodated in *Prosocheyla*, this genus is unwieldly because the included species are not close relatives. Volgin (1969) attempted to compensate for their heterogeneity by allocating them to two subgenera. One subgenus, *Prosocheyla (Prosocheyla)*, was set up for *oaklandia* Baker (type) and *traubi* Baker. The species which Smiley and Moser (1970) described as *Cheletogenes acanthus* falls into this subgenus and is possibly a close ally of *traubi*. Volgin's other subgenus, *Prosocheyla (Reckiana)*, has been assigned two species associated with scale insects on citrus trees, *buckneri* Baker (type) and *hepburni* Lawrence. The hysterosomal plating on these two species is reduced to one or two small median plates.

Undoubtly this genus is destined to be split up in the future. At present, however, the number of described species is too small to reveal how the most desirable systematic rearrangement can be achieved.

KEY TO FEMALES OF PROSOCHEYLA

1. Hysterosomal plating includes 1 median plate or shieldlike configuration of varicose striae
   centered over metapodosoma.................Subgenus *Reckiana* Volgin.................2
   Hysterosomal plating comprises 1 large shield which covers most of metapodosoma and
   opisthosoma....................Subgenus *Prosocheyla* Volgin.....................3
2. With 4 pairs of dorsolateral setae on margin of propodosomal plate ..... *hepburni* (Lawrence)
   With 1st and 2d pairs of dorsolateral setae on margin of propodosomal plate, 3rd and 4th
   pairs on independent platelets ....*buckneri* (Baker)
3. Dorsomedian setae resemble dorsolateral setae ....*oaklandia* (Baker)
   Dorsomedian setae aberrant, fragmented 4
4. Palp claw with 8 or 9 rounded teeth distributed over most of its length ....*traubi* (Baker)
   Palp claw with 3 or 4 sharply pointed teeth disposed in a compact cluster near its base
   acanthus (Smiley and Moser)

Prosocheyla oaklandia (Baker)

(Fig. 45)

Cheletogenes oaklandia Baker, 1949, Proc. U.S. Nat. Mus. 99(3238):306–307.
Prosocheyla oaklandia (Baker); Volgin, 1969, Akad. Nauk S.S.S.R., Zool. Inst., Opredel. p.
Faune S.S.S.R. no. 101:289–291.

Palp claw bears 11 to 13 blunt teeth distributed on anteromesal curvature. Outer comb much
more robust than inner comb, ca. 18 to 20 long teeth; inner comb slender, about half as long as
outer comb, ca. 20 short, delicate teeth. Palp femur appreciably bowed on lateral face, entire
sclerotized dorsal surface tuberculate. Stylophore completely covered with close-set, rounded
or spindle-shaped tubercles; protegmen elevated, forming a substantial domed hood covering
basal half of rostrum. Peritreme deeply inlaid within skeletal fold, shaped like inverted U, 4 or 5
slender links per side. Rostrum narrow, tapered to a bluntly pointed apex; protrudes about half
its length beyond protegmen; inferior adoral setae twice as long as superior adorals. With 2
substantial dorsal shields considerably separated by a wide band of tuberculate striae at humeral
sulcus; both plates decorated with heavy striae broken into bacillus-like segments, arranged in
longitudinal rows on propodosomal plate and mostly transverse rows or whorls on hysterosomal
plate; hysterosomal plate about 0.6 times as wide as propodosomal plate. Eyes present, corneas
encircled by 1 or 2 rings of broken striae. Dorsal body setae short (16–35), moderately fanlike,
15 pairs plus humerals; medians and laterals alike; 2 and 2 pairs of medians on front and rear
plates, respectively; marginals 1-2-6 not set on hysterosomal plate. Ventral body setae (4 pairs)
uniformly short, slender, simple. Normal complement of anogenital setae, 3d (dorsalmost) anals
fan-shaped.

Length ratio: leg I/idiosoma = 0.5. Setae on legs I to IV: femora 2-2-2-1, genua 3-2-2-2, tibiae
6-4-4-4, tarsi 9-8-7-7. Tarsus I lacks pretarsus and pedicel; swollen end of segment bears 6 setae,
as follows: 2 addorsals *tc*, longer than tarsal segment; 2 eupathoid paraterminals, mesal paraterminal half as long as overlying addorsal, lateral paraterminal half as long as mesal paraterminal; 2 short, fine infratarsals; solenidion *w*I (29) arises at middle of tarsal segment, long
enough to reach its distal end; guard hair minute, set on nipple with *w*; ventral seta *v* smooth,
flagelliform, arises opposite and barely exceeds length of *w*. Measurements (n = 10): length
idiosoma 316 ± 35, gnathosoma 100 ± 9, leg I 166 ± 15, tarsus I 49 ± 4.

Collection data.—California: soil under juniper, Cajon Pass, San Bernardino Co. (Uzi Nur);
soil sample, Cajon Pass, same locality (F. C. Raney); topsoil, Laguna (H. L. McKenzie); soil
sample, Encinitas (H. L. McK.); soil sample, near Oso Flaco Lake, San Luis Obispo Co. (W. H.
Lange); lichen from *Salix*, Castroville (W. H. L.); lichens, Dillon Beach, Marin Co. (F. C.
Raney); mulch of Monterey cypress, Moss Beach (W. H. Lange, E. I. Schlinger); soil sample,
Indian Hill, Sierra Co. (T. Fenner); mulch in pine forest, McKerricher Beach State Park,
Mendocino Co. (D. W. Price).

The palp femur of the male is considerably longer than that of the female and bears a strong spine on its dorsal surface. The spine originates immediately behind the dorsal seta. It resembles the palp claw in shape and curvature but is more robust. Its pointed tip is directed mesad in compressed specimens.

Prosocheyla traubi (Baker)

(Fig. 46)

Cheletogenes traubi Baker, 1949, Proc. U.S. Nat. Mus. 99 (3238) :307–308.
Prosocheyla traubi (Baker) ; Volgin, 1969, Akad. Nauk S.S.S.R., Zool. Inst., Opredel. p. Faune S.S.S.R. no. 101:291.

Palp claw with 8 or 9 teeth distributed over most of its length. Outer comb ca. 20 teeth; inner comb ca. 16 teeth. Palp femur almost straight, not much elbowed; dorsal seta on palp femur and genu fan-shaped; ventral and lateral setae on these segments acicular, smooth. Exposed part of rostrum triangular in outline, sharply pointed at apex, without projecting papillae for superior adoral setae. Protegmen and anterior part of tegmen densely stippled with microtubercles of two sizes, smaller ones more numerous than larger ones; on posterior part of tegmen, rounded microtubercles transform into short bacillus-like thickenings of broken striae. Dorsal plating weakly sclerotized, differentiated by absence of sharp striations as seen on interscutal membrane; entire dorsal cuticula plated or otherwise provided with abundant and almost uniformly spaced microtubercles; on edge view, microtubercles appear as erect rods or blunt microtrichia. Dorsal body setae: about 23 pairs plus humerals; median setae peculiarly modified. Dorsolateral setae moderately broad fans with rounded ends (30–34; humerals 40); upper ribs barbed or serrate only near attached end of blade; ribs give way to separated, upright barbs beyond middle of blade. First pair of dorsolateral hysterosomal setae much closer to midline than 2d pair; 6th pair on inferior surface, close beside fanlike posterior anals. Median setae fragmented, each consisting of many crescentic or angular sclerites (possibly modified barbs) almost equispaced on branching strands of hyaline matrix; possibly 7 pairs on propodosomal plate, 6 pairs on hysterosomal plate; anteriormost "medians" lateral to 1st "laterals" on hysterosoma. Pairs of anogenital setae: paragenitals 2, genital 2, anals 3, all but posterior anals acicular, smooth; anal cover bulbous, widened in transverse plane.

Length ratio: leg I/idiosoma = 0.6. Setae on legs I to IV: femora 2-2-2-1, tibiae 6-4-4-4, not determined for other podomeres. Tarsus I bears 4 conspicuous sensilla: mesal addorsal 90, lateral addorsal 82, mesal paraterminal 60, lateral paraterminal 40; solenidion wI 45 nearly as long as tarsal segment (50); guard seta not identified. Measurements (n = 1): length idiosoma 372 gnathosoma 140, leg I 236.

Collection data.—Holotype specimen, USNM No. 1778, collected in a tent, Stillwell Road near Ledo, Assam, North India, 1945.

The adult female is easily identified by the structure of the dorsomedian body setae, which are quite similar to those in *Paracheyletia pyriformis* (Banks).

Prosocheyla buckneri (Baker)

(Fig. 47)

Cheletogenes buckneri Baker, 1949, Proc. U.S. Nat. Mus. 99 (3238) :308–309.
Cheletogenes citrifoliatus Muma, 1964, Florida Ent. 47 (4) :241–242. New synonym.
Prosocheyla buckneri (Baker) ; Volgin, 1969, Akad. Nauk S.S.S.R., Zool. Inst., Opredel. p. Faune S.S.S.R. no. 101:293–295.

Palp claw bears 5 or 6 teeth. Outer comb 10 to 12 teeth; inner comb 8 to 10 teeth. Rostrum pointed, without rounded shoulders or projecting tubercles for superior adoral setae. Protegmen subconic, constricting to a pointed apex which merges with rostrum. Entire stylophore ornamented with delicate longitudinal striations; striae broken into tiny segments. Peritremes horseshoe-shaped, gently curved, 5 to 7 links per side. Dorsal body plating imperfectly delimited from interscutal membrane; propodosomal shield indicated by thickened, varicose longitudinal striae covering area between eyes and wide enough to incorporate 2 pairs of dorsomedian setae but not 3d and 4th pairs of dorsolateral propodosomals; a small unpaired shield between 1st and 2d pairs of dorsolateral hysterosomal setae, large enough to bear only 2 pairs of median hysterosomals. Dorsal body setae: 15 pairs plus humerals; medians may be aberrant or orthodox. Dorsolateral setae small, narrow fans or paddles with fine barbs and rounded ends; anteriormost 2 pairs of dorsolaterals appear to arise on propodosomal shield, all other dorsolaterals set on individual, trivial platelets. Dorsomedians setae amoeboid or lobulated; 2 pairs on each main shield. Posterior anal setae fanlike, set close behind 6th pair of dorsolateral hysterosomals.

Length ratio: leg I/idiosoma = 0.5. Setae on legs I to IV: femora 2-2-2-1, genua 3-2-2-2, tibiae 5-4-4-4, tarsi 8-8-7-7. Tarsus I lacks pedicel, claws, and empodia; mesal paraterminal seta 0.5 times as long as adjacent addorsal seta *tc*; solenidion *w*I overreaches end of segment, with no obvious guard seta. Tarsi II to IV with pedicels, claws, and empodia. Measurements (n = 1): length idiosoma 274, gnathosoma 94, leg I 137, tarsus I 40.

We have examined one paratype of *citrifoliatus* Muma, from citrus leaves, Weirsdale, Florida (in collection of Division of Plant Industry, Entomology Section, Florida Department of Agriculture, Gainesville) and the holotype and two paratypes of *buckneri* Baker, from lemon leaves, Santa Paula, California (Holotype 1774, USNM). From these few specimens it is known that the kind and number of median setae are variable. The specimen illustrated here (Florida) has only one and a half pairs of amoeboid setae on the hysterosoma and the "plated" area is skewed. The holotype of *buckneri* (California) has one and a half pairs of amoeboid medians on the propodosoma, whereas the two pairs of medians on its hysterosoma are orthodox in form, i.e., have the form of dorsolaterals. The two paratype females of *buckneri* (California) have amoeboid medians on both plated areas.

Prosocheyla acantha Smiley and Moser

(Fig. 48)

Prosocheyla acantha Smiley and Moser, 1970, Proc. Ent. Soc. Wash. 72(2):229–236.

Palp claw with 3 or 4 pointed teeth clustered at its base. Outer comb ca. 18 teeth; inner comb ca. 12 teeth. Tegmen and protegmen similarly ornamented with pattern of closely packed tubercles some of which become attenuate near posterior rim of stylophore. Part of each maxilli-coxal area stippled with tubercles. Peritremes shaped like inverted **U**, 6 links per side. Plating on body and appendages microtuberculate, so lightly tanned that there is little distinction between scleroized and membranous cuticle. Eyes present. Dorsal setae: 24 pairs (35–47) plus humerals; dorsal setae of two kinds. Lateral setae palmate or paddle-shaped, with transparent membranous blades and sharply outlined conical petioles; blades delicately ribbed and barbed; 4 pairs on propodosoma, 6 pairs on hysterosoma. Median setae fragmented; petioles stubby, framework appears to consist of diaphanous strands which tie together clusters of small angular or crescentic fragments; 7 pairs on propodosoma, 7 pairs on hysterosoma. With 2 pairs of paragenital setae, 2 pairs of genital setae, 3 pairs of anal setae, all smooth and flagelliform except posteriormost anals; posterior anals palmate, resembling marginals of opisthosoma.

Length ratio: leg I/idiosoma = 0.7. Setae on legs I to IV: femora 2-2-2-1, genua 3-2-2-2, tibiae 6-5-4-4, tarsi 8-8-7-7. Tarsus I lacks a pedicel; its blunt distal end bears a tuft of 6 dissimilar setae, viz., 2 very long addorsals (lateral 140, mesal 140), 2 paraterminals of intermediate lengths (lateral 47, mesal 74), 2 very small infraterminals; solenidion *w*I (66) shorter than body of tarsus, without guard seta. Distal solenidion on tibiae I and II unusually robust. Smooth claws and multirayed empodia on legs II to IV. Measurements (n = 1): length idiosoma 495, gnathosoma 156, leg I 350, tarsus I 74, vertical seta 35.

Only one other species of *Prosocheyla* has been described. It closely resembles *buckneri* as described from citrus in California by Baker and in Florida by Muma.

Prosocheyla hepburni (Lawrence). Originally described as *Cheletogenes hepburni* Lawrence, 1954, Ann. Natal. Mus. 13(1):70–71; assigned to *Prosocheyla* by Volgin, 1969. Taken from purple scale, *Lepidosaphes beckii*, on citrus, near Hamburg, Peddie District, South Africa.

Eutogenes Baker

Eutogenes Baker, 1949, Proc. U.S. Nat. Mus. 99(3238):304.

Cheyletids having a long palplike foreleg terminating in a tuft of 4 ultralong bristles, one of which is identifiable as solenidion *w*I. Palp claw toothless. Palp tarsus with 2 combs and 2 sickles. Stylophore bullet-shaped, about as wide as long; protegmen elevated, but may be abbreviated or reduced to a short crescentic swelling in front of peritremes. No eyes. With 2 shields completely covering idiosoma. Dorsal body setae short, conservative fans; medians and laterals similar in form; 5 or 6 pairs set on margins of propodosomal plate. Tarsus I lacks pretarsus and pedicel.
Type species.—Eutogenes foxi Baker; type by original designation.

KEY TO FEMALES OF EUTOGENES

1. With 12 pairs of setae on propodosomal plate (6 marginals, 6 medians) 2
   With 10 pairs of setae on propodosomal plate (6 marginals, 4 medians) 4
2. Dorsal body setae broad, rounded fans, each with a pronounced skirt; dorsal seta of palp
   femur leaflike, not a fluted rod ....*foxi* Baker
   Dorsal body setae spatulate or narrow fans having no noticeable skirts; dorsal seta of palp
   femur lanceolate to rodlike, barbed 3
3. Peritremes with a single segment on each side of midline ......*narashinoensis* Hara & Hanada
   Peritremes with 4 or 5 segments per side ....*vicinus*, n. sp.
4. With 10 pairs of dorsal setae on hysterosoma 5
   With 11 or 12 pairs of dorsal setae on hysterosoma 6
5. With 1 ventral seta on tibia I simple and smooth, the other ventral seta barbed..*frater* Volgin
   Both ventral setae on tibia I barbed ....*quadrisetatus* (Berlese)
6. With 11 pairs of dorsal body setae on hysterosoma ....*punctata* Zaher & Soliman
   With 12 pairs of dorsal body seta on hysterosoma ....*africanus* Wafa & Soliman

This key to species of *Eutogenes* represents the accumulated efforts of several authors. It has limitations, since too much reliance is placed upon the structure of tibial setae and the numbers of dorsal setae on the hysterosoma. The species are quite alike and the characters employed in the key are not very stable.

Eutogenes foxi Baker
(Fig. 49)

Eutogenes foxi Baker, 1949, Proc. U.S. Nat. Mus. 99(3238):304.

Palp claw smooth, edentate. Outer comb ca. 16 long, coarse teeth; inner comb ca. 14 very thin, short teeth. Dorsal seta on palp femur and dorsolateral seta on palp genu leaflike, with 2 or 3 apical serrations drawn into prominent, sharply pointed spines. Rim of each palp coxa projects laterally to form a pronounced ledge external to recessed articulation of palp trochanter. Protegmen elevated, rounded in front, equal to about one-fourth of stylophore length in midline; stippled with numerous equispaced tubercles. Tegmen covered with coarse, densely packed tubercles which become barlike and longitudinally oriented near its posterior supporting arch. Peritremes arch forward from midline before curving rearward, possibly as many as 5 links per side. Entire dorsum covered with 2 roughened plates; plate surface bears serpentine strands of elevated tubercles, each plate with troughlike submarginal groove or gutter on each side. Dorsal body setae numerous, 25 pairs plus humerals; small (18–29). Propodosomal plate bears 6 pairs of laterals and 6 pairs of medians; hysterosoma bears 6 pairs of laterals and 7 pairs of medians, all but 1 or 2 posteriormost pairs originating on principal plate. Dorsolaterals and dorsomedians alike in form; blade of each seta oval in outline, with sharply etched radial ribs on ventral surface; ventral (inferior) ribs project beyond margin to produce a serrated or scalloped margin; upper surface bears several incomplete rows of coarse dorsal barbs; from basal stalk arises a serrated membranous skirt which encircles junction of stalk and focal point of radial ribs; margin of skirt becomes confluent with edges of main blade and transforms into marginal rows of dorsal barbs. Anal cover circular in outline, with 3 pairs of short, simple setae; paragenitals (2 pairs) and genitals (2 pairs) comparatively long flagelliform.

Length ratio: leg I/idiosoma = 0.9. Setae on legs I to IV: femora 2-2-2-1, genua 3-2-2-2, tibiae 5-4-4-4, tarsi 6-8-7-7. Ventral seta v on tarsus I very short (12), thin, smooth, set immediately below nipple of wI; tarsus ends bluntly, with apical tuft of 4 very long sensilla identified as follows: tc lateral (140), tc mesal (144), solenidion wI (133), mesal paraterminal (94). Guard seta (51) in duplex position on nipple of wI, finely barbed. Measurements (n = 1): length idiosoma 236, gnathosoma 59, leg I 207, tarsus I 39.

Eutogenes foxi Baker is redescribed from a specimen found in citrus leaf litter, Turnbull Hammock, Mims, Florida, February 1, 1968 (M. H. Muma and F .M. Summers). This specimen appears to be conspecific with Baker's holotype, USNM 1772.

Eutogenes vicinus, new species

(Fig. 50)

Outer comb robust, ca. 12 strong blunt teeth; inner comb with slender axis, ca. 18 fragile teeth. Peritremes emerge far forward on stylophore so that protegmen appears to be short. Tegmen decorated with longitudinal striae to posterior areas of muscle attachment; striae broken into short fusiform rods near midline; these grade into more rotund tubercles toward lateral margins. Peritremes arch slightly forward from point of origin before curving rearward; each arm comprises 5 slender links; terminal link extends almost to posterior rim of stylophore. Dorsal body plating roughened with tuberculate striae formed into complex whorls; tubercles round to ovoid in outline, varying considerably in size, largest about 2 μ. Humeral setae and associated platelets displaced to pleuroventral position. Dorsal body setae (27–43) numerous: 25 pairs; blades fashioned into narrow fans with rounded ends and serrate ribs; medians and laterals alike in form; 6 setae on each lateral margin of propodosomal plate; posteriormost pair of hysterosomals displaced to venter, each seta on isolated platelet. Anal cover circular, with 3 pairs of short, simple setae; paragenitals (2 pairs) and genitals (2 pairs) relatively long, flagelliform; anterior paragenitals close beside apex of genital slit.

Length ratio: leg I/idiosoma = 0.9. Setae on legs I to IV: femora 2-2-2-1, genua 3-2-2-2, tibiae 5-4-4-4, tarsi 8-8-7-7. Tarsus I bears all its setae on or close to distal end; 4 ultralong elements identified as follows: tc lateral 199, tc mesal 203, solenidion wI (164), paraterminal (140)—1 paraterminal believed to be absent; 2 justaposed infraterminals very minute; guard seta relatively short (45), delicately barbed, arising on nipple of wI; very short ventrolateral seta probably ventral seta v of more conventional species. Measurements (n = 2): length idiosoma 361, gnathosoma 78, leg I 312, tarsus I 55.

Types.—Holotype female, from leaf mold beneath *Acer macrophyllum*, 31 miles W of Paso Robles, San Luis Obispo Co., Calif., May 12, 1961 (F. C. Raney and R. O. Schuster); deposited in UCD. Paratype female, from leaf litter *Equisetum, Umbellaria,* and *Ribes*, 7 miles E of Cambria, San Luis Obispo Co., Calif., May, 1961 (F. C. R. and R. O. S.); deposited in USNM. A sample of one cubic foot of leaf litter from each situation yielded only one specimen of *vicinus* per sample.

E. vicinus is considerably larger than *foxi* Baker, and its dorsal setae are more nearly spatulate in shape. In *vicinus* the protegmen is foreshortened and the longitudinal striae on the midsection of the tegmen are broken into fusiform rods; in *foxi* the protegmen forms a substantial conical protuberance in front of the peritremes, and the midregion of the tegmen is covered with densely packed subcircular tubercles. The spot character for separating *vicinus* from *foxi* is the form of the dorsal seta of the palp femur and the dorsolateral seta of the palp genu. These setae in *foxi* are leaflike, and each has two or three apical serrations drawn into sharply pointed spines. These setae in *vicinus* are stout, fluted rods having uniformly barbed ribs or splines.

Six additional species of *Eutogenes* have been described:

Eutogenes quadrisetatus (Berlese).
  Cheletogenes quadrisetatus Berlese, 1931, Redia 9:79. Collected in Java, Indonesia.

  Volgin (1958) transferred this species to *Eutogenes*.

Eutogenes frater Volgin, 1958, Entomol. Oboz. 37(2):460–463.
Eutogenes narashinoensis Hara and Hanada, 1960, Japanese Jour. Sanitary Zool. 11(1):25–27.
  Described as free-living, from Narashino, Chiba Prefecture, Japan.
Eutogenes citri Gerson, 1967, Acarologia 9(2):363–367. Holotype male, from lemon, Karmon, Israel.

  The female has not been identified and is not included in the key presented in this paper.

Eutogenes punctata Zaher and Soliman, 1965, Bull. Soc. Ent. Egypte 49:65–66. From palm trees, Giza, Egypt.

  In the original description the authors state that there are eleven pairs of dorsal setae on the hysterosomal shield, whereas Wafa and Soliman (1968) give ten pairs as the number on this shield. Although there is a printer's mistake in the first paper, we have taken eleven pairs to be the normal number, as illustrated in figure 1 of a paper by Zaher and Shehala, 1965, Bull. Soc. Ent. Egypte 49.68, mislabeled *Oligonychus vitis.*

Eutogenes africanus Wafa and Soliman, 1968, Acarologia 10(2):225–227. Associated with scale insects on plam trees; males identified from bird nests in Egypt.

Acaropsis Moquin-Tandon

Acaropsis Moquin-Tandon, 1863, Eléments de Zool. médicale, p. 314; Oudemans, 1906, Mém. Soc. Zool. France 19:162; Rohdendorf, 1940, Moscow Univ. Uchenye zapiski, Wissensch. Ber. 42:76–78; Baker, 1949, Proc. U.S. Nat. Mus. 99(3238):312–313; Volgin, 1955, Akad. Nauk S.S.R., Zool. Inst., Opredel. p. Faune S.S.S.R. no. 59:174.

Palp tarsus with 2 sickles and 1 orthodox comblike seta; homologue of inner comb plain. Rostrum produced to a pointed, slightly notched apex. Protegmenal part of stylophore subconic, its margins converging in front of stylet bases and its tip pressed into median gutter of rostrum. Peritremes turn rearward without abrupt change in curvature. Idiosoma incompletely covered with 2 median dorsal shields. Eyes present. Dorsal body setae acicular to lanceolate but not spatulate or fanlike; no peculiarly modified dorsomedian setae. Humeral setae noticeably longer than nearby dorsals. Guard seta of tarsus I very minute or absent. Tarsi I to IV with paired claws and multirayed empodia; claws of tarsus I may be diminutive as compared with those of tarsi II to IV.
  Type species.—Tyroglyphus mericourti Laboulbene. Monotypic.

After Oudemans (1906) revised the cheyletids, the genus *Acaropsis* was recognized as comprising three species which have a trapezoidal shield on the propodosoma and only one comblike seta on the palp tarsus. Baker (1949) erected *Chelacaropsis* for an allied species, *C. moorei*, which has no obvious dorsal plates. Then Volgin (1962) founded a kindred genus, *Neoacaropsis*, for a new species, *N. granulatus*, and he included in it a second new species (*rohdendorfi*) and one of the species (*kulagini*) formerly classified in *Acaropsis*. Gerson (1967) added *Neoacaropsis volgini*. Volgin was dissatisfied with this arrangement and, in 1969, he split *Neoacaropsis* by creating a new genus *Acaropsella* for three of these species: *rohdendorfi* (type), *kulagini*, and *volgini*. This left only the type species in *Neoacaropsis*.

Acaropsis, the original genus of this cluster of genera, now contains four species: *mericourti* (type), *rufus*, *sollers*, and *docta*. We have often recovered living specimens of *sollers* from floor debris in hay barns by processing such samples in a Tullgren apparatus.

KEY TO FEMALES OF ACAROPSIS
(Based on Volgin, 1969)

1. Dorsal body setae acicular, smooth ....*mericourti* (Laboulbene)
   Dorsal body setae lanceolate, barbed 2
2. Palp femur 1.6 to 1.7 times as long as it is wide; dorsal seta of palp femur shorter than
   femoral segment; 3d pair of medioventral body setae anterior to both pairs on coxae IV;
   paragenital setae barely longer than genitals ....*docta* (Berlese)
   Palp femur 2.0 to 2.2 times as long as it is wide; dorsal seta of palp femur longer than
   femoral segment; 3d pair of medioventral body setae aligned with setae on coxae IV;
   paragenital setae twice as long as genital setae ....*sollers* Rohdendorf

Acaropsis sollers Rohdendorf
(Fig. 51)

Acaropsis sollers Rohdendorf, 1940, Moscow Univ. Uchenye zapiski, Wissensch. Ber. 42:78–80.
Acaropsis callida Rohdendorf; Volgin, 1969, Akad. Nauk S.S.S.R., Zool. Inst., Opredel. p. Faune S.S.S.R. no. 101:311.

Palp claw with 3 or 4 pointed basal teeth. Outer comblike seta bears ca. 14 coarse teeth. Seta corresponding to inner comb of other genera is acicular, pointed, with no teeth or barbs. Superior adoral setae (35) set on upper face of rostrum without altering the almost straight-line profile of its margins; superior adorals much longer than inferior adorals. Entire stylophore ornamented with very faint, broken striae, longitudinal in direction; 6 to 8 striae form lightly elevated ridges which disappear a short distance behind peritremes. Peritremes arcuate, with 7 (6) links of uniform length and diameter. Dorsal body plates lightly sclerotized, not sharply outlined or significantly ornamented. Eyes inconspicuous, not encircled by concentric striae. Hysterosomal plate truncate in front, rounded behind, not wide enough to encompass 1st pair of dorsolateral setae on this part of body; margin of plate notched in front of 2d dorsolateral seta to accommodate a dermal pore; 2d dermal pore, in front of 3d dorsolateral seta, close to but not on plate margin. Humeral setae acicular, flagelliform, at least 110, much longer than dorsal body setae, almost imperceptibly barbed. Dorsal setae: 16 pairs plus humerals, with noticeable differences in length in setae of different pairs (27–51); each seta has a minute basal granule and a slender tapered blade, is barbed and fluted, usually with 1 or, rarely, 2 barbed ribs on convex surface; 4 marginal and 3 median pairs on propodosomal plate; 3 lateral and 3 median pairs on hysterosomal plate; 3 pairs on trivial platelets. With 3 pairs of setae on genital covers and 1 pair of paragenital setae; evidently anteriormost genital is a displaced paragenital. Anal setae simple, not bifid.

Legs comparatively slender, e.g., greatest diameter of femur I is 0.4 times its length as measured from ventral condyle to distal end. Length ratio: leg I/idiosoma = 0.7. Setae on legs I to IV:

femora 2-2-2-1, genua 3-2-2-2, tibiae 6-5-4-4, taris 9-8-7-7. Tarsus I: solenidion wI (32) arises closer to addorsal setae *tc* than to basal end of tarsus, i.e., on distal half; guard seta may be present but not observed; ventral seta *v* set on proximal half of tarsus (body), 20μ behind wI; mesal paraterminal seta on pedicel does not overreach ends of empodial raylets. Measurements (n = 5): length idiosoma 479, gnathosoma 143, leg I 313, tarsus I 106.

Collection data.—Sweepings from cattle pens, U. C. campus, Davis; feed trash, swine barns, same locality.

The authorship of *sollers* and *callida* was ascribed to B. Kuzin by Rohdendorf (1940) and by Volgin (1955 ,1969), but apparently Kuzin's original work was not independently published. We therefore show the authorship of these species to be Rohdendorf, not Kuzin. The species from California which we have identified as *sollers* Rohdendorf seems to fit the description of *docta* as given by A. M. Hughes (1961). The three other species of this genus are not known to occur in the United States:

Acaropsis mericourti (Laboulbene), 1851, Ann. Soc. Ent. France (ser. 2)9:302. Described in *Tyroglyphus*, from Newfoundland and Canada; habitat not known.

Acaropsis docta (Berlese), 1886, Acari . . . , Prostigmata, fasc. 33, no. 1 (*Cheyletus docta*); Oudemans, 1905, Ent. Ber. Nederl. Ent. Ver. 1(21):209; Oudemans, 1906, Mém. Soc. Zool. France 19:163–168.

Acaropsis rufus (Karpelles); described in Cheyletus by Karpelles, 1884, Berl. Ent. Zeitschr. 28:231; assigned to *Acaropsis* by Oudemans, 1906. Associated with beetles, Congo Region, Africa.

We are unable to accommodate this species in the key for lack of sufficient information.

Neoacaropsis Volgin

Neoacaropsis Volgin, 1962, Entomol. Oboz. 41; transl. in Entomol. Rev. 41(3):425.

Palp claw with 5 or 6 acutely pointed teeth. Inner comblike seta of palp tarsus of other genera represented here by a slender seta which bears a few fine barbs on its mesal face—not a comb in usual sense. Dorsal body setae spatulate, widened only enough to show several barbed ribs. Humeral seta similar to adjacent dorsolaterals in form and size. With 1st pair of dorsolateral hysterosomal setae implanted on anterolateral corners of hysterosomal plate. Eyes conspicuous. Tibia I with 4 setae of conventional form plus a minute solenidion. Anal setae of last pair flattened and spatulate to resemble dorsals. Each claw on tarsi II to IV with a swollen base and projecting apophysis.

Type species.—*Neoacaropsis granulatus* Volgin; by original designation.

N. granulatus Volgin, 1962, is the only representative of this genus in the present arrangement of *Acaropsis*-like species.

Acaropsella Volgin

Acaropsis Volgin, 1969, Akad. Nauk S.S.S.R., Zool. Inst., Opredel. p. Faune S.S.S.R. no. 101:314–315.

Acaropsella shares most of its features with *Neoacaropsis*. The characters which distinguish the first from the second are as follows:

Tibia I bears 5 setae of conventional type (4 in *Neoacaropsis*) plus a minute solenidion. Claws on tarsi II–IV smooth hooklets, without basal apophyses. Anal setae acicular and nude; setae of last 2 pairs of anals in *Acaropsella kulagini* and *A. rohdendorfi* bifid or forked in most specimens.

Type species.—*Neoacaropsis rohdendorfi* Volgin; by original designation.

The short, spatulate humeral setae readily distinguish species in this genus from those in *Acaropsis*. The presence of basal apophyses on the claws of tarsi II to IV is a spot character useful for separating *Neoacaropsis* from *Acaropsella;* in the latter the claws are smooth hooklets, without peculiar excrescences.

KEY TO FEMALES OF ACAROPSELLA

1. Hysterosomal plate with no marginal notch between 2d and 3d dorsolateral setae; genu II
   with both setae spatulate, not having long finely pointed tips ....*rohdendorfi* (Volgin)
   Hysterosomal plate with a marginal notch between 2d and 3d dorsolateral setae; genu II
   with 1 widened spatulate seta and 1 barbed acicular seta having a long, fine tip............2
2. Dorsal seta of palp femur lanceolate, coarsely barbed, blunt-tipped; all anal setae plain,
   not forked ....*volgini* (Gerson)
   Dorsal seta of palp femur acicular, finely barbed, drawn to a slender point; several anal
   setae forked (cf. fig. 29a) ....*kulagini* (Rohdendorf)

Acaropsella kulagini (Rohdendorf)

(Fig. 52)

Acaropsis kulagini Rohdendorf, 1940, Moscow Univ. Uchenye zapiski, Wissensch. Ber. 42:78.
Neoacaropsis kulagini (Rohdendorf); Volgin, 1962, Entomol. Oboz. 41; transl. in Entomol. Rev. 41(3):427–428.
Acaropsella kulagini (Rohdendorf); Volgin, 1969, Akad. Nauk S.S.S.R., Zool. Inst., Opredel. p. Faune S.S.S.R. no. 101:319.

Palp claw bears 3 to 5 sharply pointed teeth. Outer comb ca. 13 teeth; inner comb acicular, with numerous exquisitely fine barbs on convex border, not comblike in usual manner. Superior and inferior adoral setae of rostrum very nearly equal in length (32). Stylophore longitudinally striate, striae broken into short, bacillus-like segments; about 8 striae on tegmen developed into slightly raised ridges which blend with other striae a short distance behind peritremes. Peritremes with 5 or 6 slender links, length of links increasing progressively from 1st to rearmost. Dorsal plates clearly outlined, provided with fine, broken striae, predominantly transverse in direction. Eyes sharply delimited by a mantle of 5 or 6 concentric circles of striae. With 1st dorsolateral setae on hysterosoma arising on trivial platelets almost contiguous with anterolateral angles of median shield, isolated from it by several rows of striae in some specimens. Margin of hysterosomal plate notched between 2d and 3d dorsolateral setae; a diagonally placed, slotlike pore perforates this plate between 1st and 2d dorsolateral setae. Humeral setae identical with nearby dorsals in shape and length. Dorsal setae: 16 to 17 pairs plus humerals; no appreciable difference in length between setae of different pairs (39–47); blade of each seta narrow spatulate, greatest width about 4 times length, with 1 to 3 coarsely barbed ribs on convex surface; 4 marginal plus 3 (or 4) median pairs on propodosomal plate; only posterior pair of hysterosomals unmistakably independent of hysterosomal plate. With 1 pair of ventral setae (i.e., paragenitals) straddling front end of anogenital papilla; 2 pairs on genital covers; 3 pairs anals, 2d and 3d anals often forked or bifid.

Leg podomeres uniformly slender. Length ratio: leg I/idiosoma = 0.7. Setae on legs I to IV: femora 2-2-2-1, genua 3-2-2-2, tibiae 6-4-4-4, tarsi 9-8-7-7; minute solenidion absent on tibia II. Tarsus I: solenidion *w*I (46) originates midway between base of segment and origins of addorsals *tc*; vestigial guard seta present but unusually difficult to observe; ventral seta *v* 12μ behind *w*I; a 2d ventral (azygos) arises halfway between origins of *v* and tarsal pedicel; mesal paraterminal seta arises immediately behind apical bulb of tarsal pedicel and projects far beyond empodial raylets. Measurements (n = 10): length idiosoma 472 ± 48, gnathosoma 126 ± 6, leg I 350 ± 20, tarsus I 119 ± 5.

Collection data.—California: numerous specimens recovered from bark of grape vines, Lodi (R. O. Schuster); deep litter in hay barns, Davis (R. L. Witt).

Apart from certain minute details which could not be checked by a comparison of types, the specimens from California are so like *kulagini* Rohdendorf that the systematics of the group will be less confused if our specimens are provisionally identified as Rohdendorf's species. We have studied but one of the species of this genus. The others are:

Acaropsella rohdendorfi (Volgin). Originally described as *Neoacaropsis rohdendorfi* Volgin, 1962, Entomol. Rev. 41(3):425–428. From remnants of maize grains, Blasgodarnoye, Stavropol subdistrict, U.S.S.R.

Acaropsella volgini (Gerson). Originally described as *Neoacaropsis volgini* Gerson, 1967, Acarologia 9(2):361–363. From pine needle litter, Rehovot, Israel.

Acaropsella aegyptiaca (Wafa and Soliman). Originally described as *Acaropsis aegyptiaca* Wafa and Soliman, 1968, Acarologia 10(2):221–222. From bird nests, manure, cotton seeds, fallen citrus fruits and leaf trash; Giza and Tahreer Province, Egypt.

This must be a provisional generic assignment because the authors emphasized the characters of the species which are shared with species in *Acaropsis*. This species could not be included in the key to species of this genus.

Chelacaropsis Baker

Chelacaropsis Baker, 1949, Proc. U.S. Nat. Mus. 99(3238):315.

Baker created this genus to accommodate one species allied to *Acaropsis* but having no indication of dorsal plates. The type species has only one comblike seta on the palp tarsus. Eyes are present. The dorsal body setae are described as "broadly clavate-serrate." Two specimens on the type slide, male and female, USNM 1777, are in very poor condition. Features still evident on these specimens provide no new information about how to recognize the species.

Type species.—Chelacaropsis moorei Baker, 1949, Proc. U.S. Nat. Mus. 99(3238):315–316.

Cheyletia Haller

Cheyletia Haller, 1884, Arch. Naturg. 50(2):233–234; Oudemans, 1906, Mém. Soc. Zool. France 19:126–127; Rohdendorf, 1940, Moscow Univ. Uchenye zapiski, Wissensch. Ber. 42:90; Baker, 1949, Proc. U.S. Nat. Mus. 99(3238):297–298; Volgin, 1955, Akad. Nauk S.S.S.R., Zool. Inst., Opredel. p. Faune S.S.S.R. no. 59:169.

Our concept of the genus *Cheyletia* remains nebulous. There is, however, full accord with Volgin (1955) that its type species, *Acarus squamosus* De Geer = *Cheyletia laureata* Haller, resembles the several species now assigned to *Neoeucheyla*. Our knowledge of cheyletids is too limited to permit a clear differentiation between *Cheyletia* and *Neoeucheyla;* the key character employed here—inner sickle inflated or inner sickle acicular—may not suffice when generic concepts within the family are better defined.

The present concept of the genus rests upon Oudemans' (1897) redescription of *squamosus* De Geer, which he took from a bird, *Fringilla* (*Chlorospiza*) *chloris*. He noted that the mites described by De Geer were taken from a bug, *Aradus betulae* L. If one accepts the opinion of the first reviser of the family (Oudemans, 1906) that the mites which he found on *Fringilla* are conspecific with those which De Geer described from *Aradus*, then the type for *Cheyletia* is *squamosa* (De Geer) in the sense of Oudemans, 1897.

Volgin (1955) apparently felt that these represented two distinct species, *Acarus squamosus* De Geer from an insect and *Cheyletia squamosa* (De Geer) of Oudemans (1897, 1906) from a bird. He gave a new name, *Cheyletia papillifera*, to the latter.

We have examined the specimens which Baker (1949) illustrated as *Cheyletia squamosa* (De Geer) from Birnamwood, Wisconsin (no plant or host association indicated). There is close agreement between these specimens and the figures given by Oudemans (1897). But the examples are imperfectly preserved and some of the anatomical details vital for taxonomic discrimination are obscure. For example, the tergum appears to be plastered with large squamate setae having a fine, flocculent reticulum within their matrices. The number and arrangement of these dorsomedians cannot be discerned. Volgin attempted to cope with Baker's description by providing a new name, *Cheyletia americana*. It now appears that this is a species of dubious worth.

The species now accommodated in the genus are:

Cheyletia squamosa (De Geer)
   Acarus squamosus De Geer, 1778, Mémories . . . , 7:116.
   Cheyletia laureata Haller, 1884, Arch. Naturg. 50(2):223.
Cheyletia papillifera Volgin, 1955, Akad. Nauk S.S.S.R., Zool. Inst., Opredel. p. Faune S.S.S.R. no. 59:171. New name for *Cheyletus squamosus* Ouds. 1897 (not De Geer, 1778).
Cheyletia aradiphila Volgin, 1966a, Akad. Nauk S.S.S.R., Trudy Zool. Inst. 37:279–280.

Hypopicheyla Volgin

Hypopicheyla Volgin, 1969, Akad. Nauk S.S.S.R., Zool. Inst., Opredel. p. Faune S.S.S.R. no. 101:242–243.

The two species currently assigned to *Hypopicheyla* are very close relatives of the species contained in *Cheyletia*. Volgin based this genus on *Hypopicheyla elongata* Volgin, and he transferred from *Cheyletia* to this genus one of his older species, *H. mirabilis* (Volgin).

The palp claw of *Hypopicheyla* has numerous (8 to 11) small teeth distributed along most of its length, and there are 16 to 18 pairs of squamate dorsomedian setae equally apportioned to the two dorsal plates of females. As presently understood, females of species in *Cheyletia* have teeth restricted to the basal half of the palp claw, and about 12 pairs of squamate dorsomedian setae are about equally distributed to the two dorsal plates.

Numerous specimens which we have taken from leaf trash and topsoil in California are believed to be *H. elongata* Volgin. Volgin's types were found in association with several species of flat bugs, *Aradus* spp.

Hypopicheyla elongata Volgin
(Fig. 53)

Hypopicheyla elongata Volgin, 1969, Akad. Nauk S.S.S.R., Zool. Inst., Opredel. p. Faune S.S.S.R. no. 101:243–245.

Palp claw with 8 to 10 tiny teeth distributed over most of its length, loosely enwrapped by cupped blade of dorsal seta of tibia. Inner and outer combs alike, ca. 20 teeth on each. Outer sickle normal but very difficult to locate on most specimens. Inner sickle normal, not inflated. Cheliceral stylets originate far in front of peritremes, their basal sclerites anchored within anterior half of protegmen. Body of stylophore longitudinally striate over-all; transverse arms of peritremes emerge near its center so that lengths of protegmen and tegmen are nearly equal. Protegmen elevated, bulbous, rounded in front, covering proximal two-thirds of rostrum. Both pairs of adoral setae originate close to tip of rostrum, ventral pair set close behind lateral pair, ventrals (24) much longer than laterals (10). Arms of peritremes form a transverse girdle across midsection of stylophore, each arm comprising 3 or 4 cells in its transverse portion and possibly not more than 1 cell in posteriorly directed portion. Idiosoma wholly covered with 2 overlapping transparent plates which, in turn, are almost covered with an overlay of squamate setae; hysterosomal plate folds over tip of opisthosoma to cover an appreciable area of venter behind anal protuberance. Humeral plates extensive, displaced to ventrolateral position, interposed between coxae II and III. Eyes conspicuous, protuberant. Dorsal setae: 25 to 27 pairs including humerals. Dorsolateral setae flabellate, cycloid to obovate, cupped and ribbed, small (23–35); upper ribs represented by a few sinuous rows of blunt denticles; ribs of concave surface more sharply etched, radially disposed; 4 pairs on propodosomal plate; 5 pairs spaced far apart around periphery of hysterosomal plate; 1 or 2 of these pairs may be displaced to opisthosomal venter. Median setae squamate, numerous, variable in number, 16 to 18 pairs about equally apportioned to propodosomal and hysterosomal plates; these setae are enlarged scales, variable in outline, with minute pedicels; blades of these setae uniformly granular. Anterior setae of coxae III and 1st and 2d pairs of anal setae fanlike.

Length ratio: leg I/idiosoma = 0.6. Setae on legs I to IV: femora 2-2-2-1, genua 3-2-2-2, tibiae 5-4-4-4, tarsi 9-8-7-7. Tarsus I: solenidion wI minute (7), one length distant from base of guard seta; guard seta (74) leaflike, its broad blade tapered to bluntly rounded tip, seta 0.6 times as long as entire tarsus; ventral seta v acicular, barbed. Measurements ($n = 6$): idiosoma 295, gnatosoma 63, leg I 172, tarsus I 74.

Collection data.—From soil beneath pine, Grass Valley, Calif. (D. W. Price); vineyard soil, Lodi, Calif. (R. O. Schuster).

The other known species of *Hypopicheyla* is:

Hypopicheyla mirabilis (Volgin)
    Cheyletia mirabilis Volgin, 1955, Akad. Nauk S.S.S.R., Zool. Inst., Opredel. p. Faune S.S.S.R. no. 59:170.

According to Volgin's (1969) key, the best discernible differences between *mirabilis* and *elongata* are evident on the ventral aspect of the opisthosoma. In *elongata* the anogenital covers are displaced far forward so that the genital plates lie close behind coxae IV. The distance between the anal papilla and the hind margin of the opisthosoma exceeds the greatest width of the entire assembly of anogenital sclerites. Also the dorsal hysterosomal plate tucks under the rear part of the body, and the sixth pair of dorsolaterals arise on the ventral surface, farther forward than the fifth pair; i.e., the setae of the fifth and sixth pairs do not align in a straight crossrow. In *mirabilis* the anogenital covers are in the more usual subterminal position; the four setae customarily called dorsolaterals are set in a straight row across the underside of the opisthosoma.

Neoeucheyla Radford

Cheyletia (*Eucheyla*) Berlese, 1913, Redia 9:79.
Eucheyla Berlese; Vitzthum, 1931, Handbuch der Zoologie, p. 146; Baker, 1949, Proc. U.S. Nat. Mus. 99(3238):290–291.
Neoeucheyla Radford, 1950, *nom. nov.* for *Eucheyla* Berlese; preoccupied by *Eucheyla* Dejean et Boisduval, 1830 (Coleopt.), Radford, 1950, Int. Union Biol. Sci. (ser. C) No. 1:82; Baker and Wharton, 1952, An Introduction to Acarology, p. 233; Volgin, 1946a, Akad. Nauk S.S.S.R., Zool. Inst., Parazitol. Sbornik 22:88–89.

Bothrocheyla Volgin, 1969, Akad. Nauk S.S.S.R., Zool. Inst., Opredel. p. Faune S.S.S.R. no. 101: 261–262. New synonym.

Inner sickle on palp tarsus inflated. Elevated protegmen comprises a prominent part of stylophore, expanded hoodlike over basal two-thirds of rostrum, its front margin truncate or rounded. Subcapitular setae present. Eyes present. With 2 plates covering entire dorsum of idiosoma; humeral plates displaced to pleuroventral position. Dorsolateral setae flabellate, cycloid to obovate, fluted and crossveined. Median setae numerous, some or all squamate and reticulate; in slide mounts these are appressed to dorsum, almost covering dorsal plates. At least 1 pair of anal setae fanlike. Guard seta on tarsus I leaflike, overlying small solenidion *w*I; guard seta arises a short distance behind *w*I. Tarsi with normal claws and empodia.

Type species.—Cheyletia (Eucheyla) loricata Berlese; by original designation.

Volgin (1969) recognized the diversity of species in *Neoeucheyla*, and he partitioned the genus into two subgenera, *N.* (*Neoeucheyla*) and *N.* (*Cunliffella*). We believe that this action was strategic but also that the differences between the two groups of species are great enough to justify a complete split in the genus. Accordingly, the second subgenus, *N.* (*Cunliffella*), is accorded full generic status in this paper.

On the other hand, the creation of another genus, *Bothrocheyla* Volgin, for *Neoeucheyla pavlovskyi* does not seem to accomplish an equally serviceable end. The relegation of *pavlovskyi* to a separate genus because its several pairs of dorsomedian setae are unlike is inconsistent with other, related decisions, e.g., the assignment of *Cunliffella whartoni* and *C. panamensis* to the same genus. These two species appear to be compatible congeners, yet *whartoni* has dorsomedians all the same kind whereas *panamensis* has dorsomedians of two kinds. A priori, *Bothrocheyla pavlovskyi* Volgin is herewith reassigned to its parent genus.

Key to Females of Neoeucheyla

1. Dorsomedian setae of two kinds, some decumbent squames, others resembling dorsolaterals.. 2
    Dorsomedian setae uniformly of one kind 3
2. Dorsal setae on palp femur forked, with 5 or 6 prongs ....*typhosa*, n. sp.
    Dorsal setae on palp femur fan-shaped, entire .... *pavlovskyi* Volgin
3. Dorsomedian setae flattened into overlapping, decumbent squames 4
    All dorsomedian setae resemble dorsolateral setae in form and elevation 5
4. About 9 pairs of dorsomedian setae on hysterosomal plate .... *bulgarica* (Volgin)
    About 5 pairs of dorsomedian setae on hysterosomal plate .... *loricata* (Berlese)
5. Dorsomedian setae very large in diameter, each considerably overlapping the one next behind
                                    ornata Wafa and Soliman
    Dorsomedian setae of modest size, such that right and left members of same pair or members
    of successive pairs mostly do not overlap................................*mumai* Volgin

Neoeucheyla typhosa, new species

(Fig. 54)

Palp claw gently bowed, with 1 pointed tooth at midpoint of its length. Inner and outer comblike setae bear a small number of coarse teeth, possibly 10 to 12 teeth on each. Inner sickle faintly S-shaped, inflated. Outer sickle forked or bifid near distal end. Dorsal setae on palp tibia stubby, brushlike, with abrupt basal bend, basal bulb of shaft and alveolus unusually enlarged. Dorsal and dorsolateral setae on palp femur peculiarly modified; blade of each seta forklike, comprising 5 or 6 branched spikelets. Cheliceral stylets very long, unusually slender. Transverse arms of peritremes divide body of stylophore into almost equal anterior and posterior sections; anterior section—the protegmen—elevated, concave on front margin; forms enlarged hood overlying proximal half of rostrum, its cuticula faintly striated lengthwise; tegmental section delicately sclerotized, with sharply etched longitudinal striae. Rostrum with straight margins, slightly broader at base than apex; apex rounded between lateral adoral setae; ventral adoral setae far behind lateral adorals, their alveoli below front edge of protegmen. With 1 pair of subcapitular setae, in normal position. Idiosoma entirely covered by 2 dorsal plates; plate surfaces lightly striated; humeral plates and setae displaced to pleuroventral position. One pair of eyes. Dorsal setae: 26 pairs excluding humerals; 16 pairs flabellate, 10 pairs squamate. Flabellate setae cycloid, cupped, ribbed on both surfaces: 8 to 10 coarsely toothed, thick ribs on convex surface and about twice as many thin-lined ribs on concave surface; ribs of both sides interlaced with meshwork of fine crossveins. All dorsolateral setae flabellate, anteriormost pair (i.e., verticals) about twice as long (62) as those on rear margin of opisthosomal plate (30). With 6 pairs of dorsomedians (3 pairs on propodosoma, 3 pairs on hysterosoma), flabellate in form, about equal in size. Squamate setae are thin plates with a fenestrated supporting framework; 4 pairs on propodosoma (3 pairs placed sublaterally, 1 pair in line with other dorsomedians); 6 pairs on hysterosoma (4 pairs sublaterals, 2 pairs aligned with other dorsomedians). Ventral setae of body proper acicular, smooth, relatively short; anterior seta on coxa III spatulate, broad but pointed; posteriormost pair of anal setae flabellate.

Length ratio: leg I/idiosoma = 0.7. Setae on legs I to IV: femora 2-2-2-1, genua 3-2-2-2, tibiae 5-4-4-4, tarsi 9-8-7-7. Tarsus I: solenidion *w*I short (18), slender, situated in front of guard seta by half its length; guard seta foliaceous, ribbed, about 0.7 times as long as entire tarsus; ventral seta *v* a pointed blade fringed with long barbs. Measurements (holotype): length idiosoma 304, gnathosoma 90, leg I 204, tarsus I 90.

Collection data.—Taken from soil under pine trees, Lake Pillsbury, Napa Co., Calif. (D. W. Price).

Among the cheyletids having clamshell or flabellate dorsolateral setae and numerous dorsomedians of the squamate type, this new species appears to be the closest described relative to the type species of *Eucheyla, E. loricata* Berl. Although these two species show many differences in fine details, we believe that *typhosa* is a suitable congener. Our knowledge of *loricata* is based entirely upon Berlese's description and illustrations (Berlese, 1913, figs. 7*a–d*). His illustrations are quite clear on most points, but the sketch of tarsus I (fig. 7*a*) is probably a mislabeled representation of a terminal segment of one of the hind legs, possibly tarsus IV.

A few characters adequately distinguish *typhosa* from *loricata*. In the females of *loricata* there are five pairs of flabellate setae on the margin of the propodosomal plate, and all the medians on the dorsum are modified or squamate. In the females of *typhosa* there are only four pairs of lateral propodosomals but sixteen pairs of medians; six pairs of medians are flabellate and ten pairs are squamate. On the gnathosoma of *loricata* the dorsal femoral setae are spatulate or fanlike, the outer sickle is normal, and the inner sickle consists of a bulbous swelling on a slender peduncle. The gnathosoma of *typhosa* bears skeletonized forklike setae on the upper aspect of the palp femur, the outer sickle has a split end, and the inner sickle is a sigmoid strap or swollen rod.

The male of *typhosa* is much like its female counterpart in gnathosomal and propodosomal organization. A noticeable difference between the sexes is the absence of the ten pairs of squamate setae, or their homologues, on the dorsum of the male.

Neoeucheyla presently accommodates five other species:

Neoeucheyla loricata (Berlese), 1913, Redia 9:79–80.

The presence of eyes is not mentioned in Berlese's original description of this, the type species, and his illustrations are not helpful on this point. The species subsequently assigned to the genus have eyes.

Neoeucheyla bulgarica (Volgin), 1955, Akad. Nauk S.S.S.R., Zool. Inst., Opredel. p. Faune
 S.S.S.R. no. 59:171; Volgin, 1964a, Akad. Nauk S.S.S.R., Zool. Inst., Parazitol. Sbornik
 22:89–90.
Neoeucheyla pavlovskyi Volgin, 1964a, *ibid.*: 94–98.
Neoeucheyla mumai Volgin, 1969, Akad. Nauk S.S.S.R., Zool. Inst., Opredel. p. Faune S.S.S.R.
 no. 101:254–255.
Neoeucheyla ornata Wafa and Soliman, 1968, Acarologia 10(2):228–229.

Cunliffella Volgin

Neoeucheyla (*Cunliffella*), Volgin, 1969, Akad. Nauk S.S.S.R., Zool. Inst., Opredel. p. Faune
 S.S.S.R. no. 101:255.

Palp femora massive, drawn together beneath stylophore; usual position of rostrum at apex
of basis capituli disarranged by mesal encroachment of basal joints of palps. Distal palp segments compactly organized; claw short, down-curved, toothed on entire inner margin. Inner
sickle inflated. Cheliceral tegmen bilobed or cleft in front, with a pair of submarginal fossae or
vesicular chambers; protegmen not distinguishable. Subcapitular setae absent. Two plates cover
entire dorsum of idiosoma; humeral plates displaced to pleuroventral position. Eyes present.
Dorsolateral and some dorsomedian setae large, flabellate; some or all dorsomedian setae aberrant, squamate, nonbarbed. With 1 pair of anal setae flabellate. Tarsi with normal claws and
empodia; guard seta of tarsus I long, leaflike, arising considerably behind very small solenidion
wI.

Type species.—Neoeucheyla tuberculicoxa Volgin, 1964; by original designation.

Baker (1949) placed two very ornate species, *panamensis* and *whartoni*, in the
genus then called *Eucheyla*. Their relatedness to its type species, *loricata* Berlese,
appears to be somewhat tenuous, especially as regards gnathal organization, and it
seems desirable to have a new genus to accommodate them. According to Baker's
original descriptions, *panamensis* and *whartoni* appeared to be very dissimilar
creatures. With newer knowledge of cheyletid structures and improved optics, it
is now clear that these two species are closely allied congeners.

Volgin (1969) established *N.* (*Cunliffella*) as a subgenus to accommodate
Neoeucheyla tuberculicoxa (type) and the two species described by Baker. We
have elevated *Cunliffella* to generic rank in the belief that the included species are
not congeneric with those which Volgin left in *Neoeucheyla* (s. str.): *loricata*,
bulgarica, and *mumai*.

KEY TO FEMALES OF CUNLIFFELLA

1. Dorsomedian setae of two kinds, some resembling dorsolaterals, others very flattened, decumbent squames .... *panamensis* (Baker)
   Dorsomedian setae all alike, so overlapped that their separate outlines and pedicels are
   obscure 2
2. With 5 or 6 teeth on palp claw .... *tuberculicoxa* Volgin
   With 4 teeth (possibly) on palp claw .... *whartoni* (Baker)

Cunliffella panamensis (Baker)
(Fig. 55)

Eucheyla panamensis Baker, 1949, Proc. U.S. Nat. Mus. 99(3238):291.
Neoeucheyla (*Cunliffella*) *panamensis* (Baker); Volgin, 1969, Akad. Nauk S.S.S.R., Zool. Inst.,
 Opredel. p. Faune S.S.S.R. no. 101:260–261.

Gnathosoma uniquely organized. Palp femora greatly distended, so extensively articulated to
basis capituli that that their mesal rims almost meet in midventral line. Distal palp segments

compressed into compact endpiece bearing claw and combs on its apex. Palp claw sharply down-curved, splayed rather than unciform, dental ridge on its anterior face, with about 9 coarse teeth directed forward. Outer and inner comblike setae similarly fashioned, 8 to 10 teeth each. Outer sickle-like seta very slender, difficult to locate; inner sickle inflated, claviform. Rostrum diminutive, displaced upward by encroachment of palpi. Subcapitular setae absent (1 pair normal in allied genera). Tegmenal part of stylophore cleft or bilobed on anterior face; no obvious protegmen. Cheliceral stylets anchor behind transverse arms of peritremes. Peritremes arise in cleft of tegmen, transverse arms circumscribe its 2 lobes, descending arms make an anastomosing loop around a pair of peculiar submarginal fossae or vesicles; number of links indeterminate. With 2 plain dorsal plates covering entire dorsum, the margins juxtaposed at humeral sulcus. Humeral plates displaced ventrad, accommodated between coxae II and III. One pair of well-defined eyes. Dorsal setae of body and appendages elaborate, foliate; probably 24 pairs plus humerals on idiosoma. Dorsolateral setae flabellate (clamshell), irregularly spaced around margins of plating, 4 pairs on propodosoma, 6 pairs on hysterosoma, with appreciable size gradation from largest propodosomals (74) to smallest opisthosomals (35). Each dorsolateral seta has a very short peduncle to support a large suboval, cupped and fluted blade; 7 to 10 coarsely serrate or barbed ribs comprise a somewhat asymmetric framework on convex surface of blade; these are laced together with numerous crossbraces; concave surface of blade also ribbed or fluted; inferior ribs slender, raylike, nonbarbed, not coinciding with upper ribs in number or pattern. Two structural types of median dorsal setae. Arrangement of median setae varies considerably within populations; these setae may be roughly but not assuredly grouped (typed) as dorsosublaterals and dorsomedians; basic arrangement probably is 3 pairs dorsosub-laterals plus 3 pairs dorsomedians on propodosoma, 4 pairs dorsosublaterals plus 4 pairs dorsome-dians on hysterosoma. Setae of dorsomedian group tend to resemble dorsolaterals in form and size; setae of dorsosublateral group aberrant, squamate, nonbarbed, with ribs and crossbraces forming a fenestrated reticulum overlying a thin hyaline matrix. Specimens identified as deu-tonymphs have no dorsosublaterals; dorsomedian setae of nymphs may become squamate setae in adults so that squamate or transitional types replace some flabellate dorsomedians on one side or both. With 3 pairs of setae on anal papilla: 2 pairs acicular, smooth; 1 pair flabellate (fig. 55*d*). Length ratios: leg I/idiosoma = 0.6, I/II = 1.1; distribution of setae on legs I to IV: femora 2-2-2-1, genua 3-2-2-2, tibiae 5-4-4-4, tarsi 9-8-7-7. Coxa I has blunt apophysis on ventro-lateral face. All tarsi bear normal claws and empodia. Tarsus I: solenidion wI delicate, short (15), cylindrical, separated by its own length from guard seta; guard seta (69) foliate, about 0.7 times as long as tarsus, makes a canopy over wI; ventral seta v a pointed blade, both edges fringed with long barbs. Measurement (n = 10): length idiosoma 383 ± 21, gnathosoma 76 ± 4, leg I 238 ± 12, tarsus I 99 ± 5.

Collection data.—Soil humus, Silverton, Colorado (D. W. Price); humus, 3 miles SSE of Jackson, Calif.; pine leaf litter, Boyce Thompson Institute Research Sta., Grass Valley, Calif. (D. W. Price); willow leaf litter, U. C. campus, Davis, Calif. (D. W. Price); beneath log, Tomales Bay State Park, Marin Co., Calif. (M. Irwin, D. Cavagnaro).

Cunliffella whartoni (Baker)

(Fig. 56)

Eucheyla whartoni Baker, 1949, Proc. U.S. Nat. Mus. 99(3238):292.
Neoeucheyla (Cunliffella) whartoni (Baker); Volgin, 1969, Akad. Nauk S.S.S.R., Zool. Inst., Opredel. p. Faune S.S.S.R. no. 101:255–256.

Gnathosoma compact, subterminal, underlying front of propodosoma. Palp femora pulled together beneath basis capituli, their mesal rims juxtaposed. Rostrum short, displaced upward, almost surrounded by palps and stylophore. Distal palp segments shortened, almost telescoped together; palp claw toothed on entire anterior face, possibly not more than 4 splayed teeth. Outer comb ca. 10 teeth; both comblike setae stubby, hook-shaped; other sensilla on palp tarsus not clearly visible on type specimens. Stylophore widened, abruptly rounded in front, with peritremes inlaid across its foremost boundary as viewed in profile. Peritremes curve rearward then reverse direction to loop around a pair of submarginal vesicles or fossae. Protegmenal part of stylophore not apparent. With 2 extensive dorsal plates revealed by overlapped edges and arrangement of striae in transverse sulcus. Dorsal body setae of two kinds. Dorsolateral setae clamshell-like or flabellate, cupped, deeply furrowed on convex side; 10 pairs plus humerals, set into sidewalls and upturned, reflected onto back. Dorsomedian setae strongly squamate, matrices coarsely reticulate; setae overlap each other, closely appressed to dorsum; number of squames not determined, probably not in excess of 18 pairs. Genitalia and appendages resemble corresponding parts of *panamensis*.

Illustrated specimen sketched from holotype and paratypes on type slide, USNM 1767; habitat unknown, Birnamwood, Wisconsin.

The entire dorsum of *whartoni* is shingled with squamous median setae, whereas in *panamensis* some of the dorsomedian setae retain the clamshell form of the dorsolaterals and the reticulate squames do not entirely cover the tergum.

The third species in *Cunliffella* is:

Cunliffella tuberculicoxa Volgin, 1964a, Akad. Nauk S.S.S.R., Zool. Inst., Parazitol. Sbornik 22:88–97.

A restudy of Baker's species suggests the possibility that *whartoni* and *tuberculicoxa* may be synonyms. Volgin says that *tuberculicoxa* has five or six teeth on the palp claw, whereas our account of *whartoni* indicates possibly not more than four teeth. The discrepancy may be a normal variation or a miscount because the mouthparts are obscured by overlying setae or other body structures.

Samsinakia Volgin

Samsinakia Volgin, 1965, Akad. Nauk S.S.S.R., Trudy Zool. Inst. 35:296–297.

Gnathosoma arises on venter of body, partly covered by overhanging propodosoma; somewhat uropodoid in general appearance. Palp claw with 3 basal teeth. Palp tarsus bears 2 sickles, 2 combs. Dorsum completely covered with 2 extensive overlapped plates. Humeral seta displaced to ventrolateral position. Eyes present. Marginal body setae acicular, relatively short, pointed, finely barbed; 4 + 6 pairs on propodosoma and hysterosoma, respectively. Median setae aberrant, squamous, stippled with minute discs of granules; unusually numerous, overlapped, 52 pairs (24 + 28) completely covering dorsum. Posterior anal setae fan-shaped. Guard seta on tarsus I foliaceous. Claws on all tarsi.

Type species.—*Cheletophyes theodoridis* Samšiňák, 1959. Monotypic.

The generic definition is paraphrased and condensed from Volgin's (1965) presentation. The only known species, *S. theodoridis*, has several peculiar features by which it should be recognized easily. The gnathosoma is partly covered by an overhang of the propodosoma. The dorsal body setae represent a most unusual combination of types: acicular dorsolaterals in company with squamous dorsomedians; also, the large number of median setae on both plates (52 pairs) is not encountered elsewhere among cheyletids. The type species has been redescribed by Volgin:

Samsinakia theodoridis (Samšiňák), 1959, Mém. Inst. Sci. Madagascar 13A:81–83; Volgin, 1965, Akad. Nauk S.S.S.R., Trudy Zool. Inst. 35:298–299. Types recovered from beetles, *Selinus abacoides* Fairmoire, Tananarive, Madagascar.

Microcheyla Volgin

Microcheyla Volgin, 1966a, Akad. Nauk S.S.S.R., Trudy Zool. Inst. 37:279–280.

Palp claw multidentate, similar to outer comb in form and size. Inner comb scimitar-like, apparently without teeth. Peritremes abbreviated. With 2 delicately sclerotized shields covering entire tergum. Humeral setae mounted pleuroventrally. Eyes present. Dorsolateral setae flabellate (clamshell); dorsomedian setae markedly squamous. With 1 pair of genital setae. Legs short, partly ensheathed with foliaceous setae. Addorsal setae *tc* inconspicuous on all tarsi. Pretarsi I to IV have rayed empodia but lack true claws.

Type species.—Microcheyla parvula Volgin; by original designation.

Microcheyla parvula Volgin

(Fig. 57)

Microcheyla parvula Volgin, 1966*a*, Akad. Nauk S.S.S.R., Trudy Zool. Inst. 37:278–279.

Palp claw comblike, bears numerous slender teeth. Outer comb resembles palp claw; inner comb scimitar-like, apparently without teeth. Dorsal seta of palp tibia leaflike, cupped around palp claw. Palp femur carries only 3 setae, all with broad net-veined blades. Tegmenal part of stylophore somewhat globate, delicately striated; protegmen projects in front of peritremes as a narrow bullet-shaped lobe. Rostrum pointed, superior adoral setae set on prominently projecting nipples. Peritremes short, probably 3 links per side. With 2 striated shields covering entire dorsum. Humeral setae displaced to body sidewalls. General integument with wide-spaced striae so that it appears to be wrinkled. Eyes present, corneas almost hemispherical. With 10 pairs of dorsolateral setae and 3 pairs of dorsomedian setae arranged 4 + 1 and 6 + 2 on front and rear plates, respectively. Dorsolaterals flabellate (clamshell-like), cupped, peduncles notched into basal margin of almost circular blades; a few setae wider than long (e.g., verticals); blades with multibranched radial ribs and numerous pebble-like tuberosities believed to be modified barbs; dorsolaterals diminish in size in rearward progression, beginning with largest in front (verticals); sockets of these setae extraordinarily large. Dorsomedian setae extremely flattened, extensive in area, variable in shape, deeply notched for peduncles; matrix with scattered flocculent granules. Anogenital setae very short: 1 pair of paragenitals, 1 pair of genitals, 3 pairs of anals one of which is flabellate.

Length ratio: leg I/*idiosoma* = 0.5. Setae on legs I to IV: femora 2-2-2-1, genua 2-2-2-2, tibiae 6-4-4-4, tarsi 8-4-4-4. Tarsi I to IV have stubby pedicels and multirayed empodia but no claws. Guard seta of tarsus I (3) foliaceous, net-veined, almost as long as body of tarsus, forming elongate canopy over minute solenidion *w*I (4). Tarsus II lacks solenidion *w*II. Measurements (n = 10): length idiosoma 207 ± 9, gnathosoma 52 ± 1.5, leg I 102 ± 2, tarsus I 39 ± 1.2.

Collection data.—Soil and leaf trash under juniper, Cajon Pass, San Bernardino Co., Calif.

Two useful spot characters for recognizing this interesting species are the presence of empodia, the absence of claws on all tarsi and, tarsi II to IV bearing only 4-4-4 setae.

Chelonotus Berlese

Chelonotus Berlese, 1893, Acari . . . , Ordo Prostigmata, p. 77.
Chelenotus Trouessart; Oudemans, 1906, Mém. Soc. Zool. France 19:158–159.
Chelonotus Berlese; Baker, 1949, Proc. U.S. Nat. Mus. 99(3238):310–311.

Gnathosoma compact; palps stubby, segments large in girth, short in length, and may be overlaid by propodosoma. Palp claw with 1 large basal tooth. Apical sensilla on palp tarsus as follows: 1 comb with 6 to 11 teeth (usually 6 to 7), 1 enlarged lanceolate sensillum, 2 sickle-like elements, and a slender sensory peg. Peritremes ʌ-shaped, sharply reflexed into descending arms in sidewalls of chelicerae. With 2 well-sclerotized plates completely investing dorsum of idiosoma. Eyes absent. Dorsal setae relatively short, smooth or finely barbed, none ultralong; 4 pairs of laterals on propodosoma, 5 pairs on hysterosoma; no median setae present. Coxae III and IV juxtaposed, close behind coxae I and II. Pretarsi I to IV equipped with paired claws and rayed empodia.

Type species.—Chelenotus selenirhynchus Berlese, 1893, Acari . . . , Ordo Prostigmata, p. 77. Monotypic. From a squirrel, *Baginia tenuis lowii* (= *Sciurus lowii*), Borneo; from squirrels, Malaya, Borneo, and the Celebes; from a tree shrew, *Tupaia glis*, Malaya.

The genus is reviewed by Domrow (1960), who considers *Chelonotus oudemansi* Baker and *C. ewingi* Baker (1949) to be synonyms of Berlese's *C. selenirhynchus.* Thus the genus contains only one known species.

Cheletopsis Oudemans

Cheletopsis Oudemans, 1904*b*, Ent. Ber. Nederl. Ent. Ver. 1(18):163.

Palp claw with 1 or 2 teeth. Palp tarsus with 1 well-defined comb, 1 plain seta, and 2 sickle-like setae. With 1 substantial median plate incompletely covering propodosoma; no unpaired plates on hysterosoma. Eyes absent. Dorsal body setae filiform, finely barbed, 2 or more pairs ultralong. Tarsi I to IV with paired claws and rayed empodia. Currently known species are from bird plumage.

Type species.—Cheyletus nörneri Poppe, 1888, Abh. Naturw. Ver. Bremen 10:239; by original designation.

Five additional species are known:

C. impavida Oudemans, 1904*d*, Ent. Ber. Nederl. Ent. Ver. 1(19):170.
C. basilica Oudemans, 1904*d*, *ibid.*
C. animosa Oudemans, 1904*d*, *ibid.*
C. magnanima Oudemans, 1904*d*, *ibid.*
C. anax Oudemans, 1904*d*, *ibid.*

A key to the above-listed species was developed by Oudemans (1906) and augmented by Baker (1949). All are redescribed and illustrated in Oudemans' 1906 monograph. *Cheletopsis major* Oudemans, 1904, was removed from the genus by Volgin (1969).

Eucheletopsis Volgin

Eucheletopsis Volgin, 1969, Akad. Nauk S.S.S.R., Zool. Inst., Opredel. p. Faune S.S.S.R. no. 101:356.

Palp tarsus with 3 plain and 1 comblike seta. Palp claw with 1 tooth. Propodosomal shield bears 3 pairs anterolateral setae; 4th pair, 1 pair medians, and 2 humeral setae set off propodosomal plate. All dorsal propodosomal setae long and smooth. Eyes absent. Without hysterosomal shield. Venter bears a small triangular sternal plate between coxae I. Genital opening small, near posterior margin of idiosoma. Legs relatively long, with 4th pair longest. Coxae IV separated from coxae III, situated near rear edge of idiosoma, with 1 pair of setae. Trochanter IV without setae. Tarsus I without solenidion *w*I or guard seta. Dorsal setae near claws on all tarsi are flat, gradually widening to a fringed tip.

Type species.—Cheletopsis major Oudemans, 1904*b*, Ent. Ber. Nederl. Ent. Ver. 1(18):163; by original designation. From a bird, *Hemiprocne* (= *Dendrochelidon*) *mystacea* (Less.), New Guinea. Male unknown.

This monotypic genus is closely allied to *Cheletopsis* Oudemans. According to Volgin (1969), *Eucheletopsis* may be distinguished from *Cheletopsis* and other related genera on the basis of coxae IV, which is separated from coxae III, the presence of only one seta on coxae IV, and the absence of solenidion *w*I on tarsus I. In addition, the chaetotaxy of the propodosomal shield and the absence of setae on trochanter IV are diagnostic.

Cheletoides Oudemans

Cheletoides Oudemans, 1904*a*, Ent. Ber. Nederl. Ent. Ver. 1(17):154.

Peritremes arc-shaped, protegmen prominent, subconic, rostrum and adoral setae well displayed in front. Palp claw with 1 prominent basal tooth. Palptarsal sensilla simple, 2 sickle-shaped and 2 plain. Idiosoma somewhat slender, width less than half its length. Dorsal plating restricted to propodosoma; this plate an inverted pentagon, incompletely covering this body section. Eyes absent. Dorsal setae filiform, minutely barbed, many excessively long; 3 posteriormost pairs of dorsals ultralong, subequal, clustered on suranal area. All ventral setae comparatively short. Coxae III and IV juxtaposed. Solenidion wI very long, with tiny guard hair sharing its tubercle. Pretarsi I to IV normal, each with paired claws and a rayed empodium.

Type species.—*Syringophilus uncinatus* Heller, 1880, Die Naturkrafte 30:188; by original designation.

No other species are known. The type, *Cheletoides uncinata* (Heller), was described from a bird, *Pavo cristatus*, Germany. An extensive description of its immature and mature stages is given by Oudemans (1906).

Alliea Yunker

Alliea Yunker, 1960, Proc. Helminth. Soc. Wash. 27(3):278.

Peritremes arc-shaped, apparently unsegmented. Protegmen elevated, broadened into a shelf-like process, with projecting lateral lobes, almost truncate across elevated transverse fold. Palp claw without teeth. No comblike apical setae on palp tarsus; apical tuft comprises 1 smooth seta about as long as claw, 1 short, inflated, but pointed seta, 2 which correspond to sickles of other species. Idiosoma entirely covered by 2 major plates. Eyes absent. Dorsal setae numerous, short, squamiform or fan-shaped, laterals and medians alike; 16 pairs on propodosomal plate, 15 pairs on hysterosomal plate. Tarsus I with spatulate guard seta remote from wI. Tarsi I to IV with paired claws and rayed empodia.

Type species.—*Alliea laruei* Yunker, 1960, Proc. Helminth. Soc. Wash. 27(3):278–281; by original designation. From *Rattus norvegicus*, Florida.

The gnathosomal features indicated above were described from the holotype male; a gnathosoma was missing on the single allotype female. No other species are known.

Caudacheles Gerson

Caudacheles Gerson, 1968, Acarologia 10(4):645.

Palp claw smooth, toothless. Only 1 sensillum on palp tarsus comblike; other elements include 2 sickles and 1 plain seta corresponding to inner comb of other genera. Eyes absent. With 2 plates entirely covering dorsum of idiosoma, both with emphatic reticulation. Dorsal body setae cycloid or clamshell-like in form, very numerous (40 pairs); medians and laterals similar. Tarsi I to IV have paired claws and rayed empodia. Solenidion wI arises near distal end of tarsus I, close behind paired addorsal setae; foliate guard seta near proximal end of this podomere, very remote from wI. Genital and anal apertures widely separated; genital covers conventional in form and location; anal vent at apex of a posteriorly projecting lobe of opisthosoma.

Type species.—*Caudacheles khayae* Gerson, 1968, Acarologia 10(4):645–649. Monotypic. Known only from one collection, on bark of *Khaya nyasica* (Meliaceae), Rehovot, Israel.

This mite has three peculiar characters: the anus positioned on a short, rearwardly projecting lobule of the opisthosoma; the large number of dorsal body setae (40 pairs), many of them around the periphery of the dorsal plates; and the carriage of the solenidion wI near the distal end of tarsus I.

Although the external anatomy of its anal and genital parts is unique among cheyletids, this species shows a small degree of kinship with *Alliea* Yunker.

Nihelia Domrow and Baker

Hemicheyletus Lawrence, 1954, Ann. Natal Mus. 13(1):74. Unavailable under Art. 13(*b*). Type species not designated.

Nihelia Domrow and Baker, 1960, Stud. Inst. Med. Res. Malaya 29:194.
Hemicheyletus Lawrence, 1961, Publ. S. Afr. Inst. Med. Res. 11:113.
Nihelia Domrow and Baker, 1963, Acarologia 5(2):225.
Hemicheyletus Volgin, 1969, Akad. Nauk S.S.S.R., Zool. Inst., Opredel. p. Faune S.S.S.R. no. 101:378.

Palp tarsus much reduced; may be a tiny sessile nodule with 2 to 4 simple setae, none distinguishable as combs or sickles. Palp claw very large, edentate; or obscure, not readily discernible. Palps appear to comprise only 2 movable segments: femur and tibia; 1 or both podomeres may bear median or lateral apophyses; or palps may be diminutive and obscured by other specialized mouthparts. Peritremes somewhat arc-shaped, chambered; segments of transverse arms much smaller than those of decending limbs. Two major plates cover most of idiosoma; 5 to 8 pairs of setae originate on each plate. Eyes absent. Body and appendicular setae acicular, smooth or finely barbed; none of body setae appreciably longer than others; peculiar discoidal or squamous median setae occur in only one species (*N. squamosa*). Legs with coxae arranged in two groups; tarsi I to IV have paired claws and rayed empodia.

Type species.—*Nihelia calcarata* Domrow and Baker, 1960. Stud. Inst. Med. Res. Malaya 29:194–197; by original designation. From *Herpestes* (mongoose), Thailand.

Other described species are:

N. curvidens (Lawrence), 1948, Parasitology 39:40; Lawrence, 1954, Ann. Natal Mus. 13(1):74. From fur of slender mongoose, *Herpestes sanguineus*, Natal and Transvaal, Africa.

N. lemuricola (Lawrence), 1948, Parasitology 39:39. From thick-tailed bushbaby, *Galago crassicaudatus*, Natal, Africa.

N. squamosa Domrow and Baker, 1963, Acarologia 5(2):227–229. From a Squirrel, *Menetes* sp., Nan Province, Thailand. Volgin (1969) assigned this species to a new subgenus, *Sciurocheyla*.

This genus comprises a heterogeneous assemblage of species, each having peculiar modifications of the generalized cheyletid organization. These mites are known only by specimens taken from the fur of small mammals in southern Asia and Africa.

Volgin (1969) considers *Nihelia* Domrow and Baker (1960) to be a synonym of *Hemicheyletus* Lawrence (1954). Domrow and Baker (1963) and the present authors consider *Hemicheyletus* Lawrence (1954) to be an invalid name, and *Hemicheyletus* Lawrence (1961) to be a synonym of *Nihelia* Domrow and Baker (1960). The problem arises because Lawrence (1954) did not designate a type for his genus *Hemicheyletus* from the two species originally included, *H. lemuricola* and *H. curvidens*. According to the International Rules, Art 13(b), a genus-group name must be accompanied by the definite fixation of a type species. Domrow and Baker (1963), by a strict interpretation of this rule, declared that *Hemicheyletus* Lawrence (1954) was an invalid name. The first designation of a type for the genus-group which currently includes these four species was *Nihelia calcarata* Domrow and Baker (1960). The following year, Lawrence (1961) designated *H. curvidens* (Lawrence) (= *Cheletiella curvidens* Lawr., 1948) as the type of *Hemicheyletus*. However, since Domrow and Baker (1963) included Lawrence's two species, *lemuricola* and *curvidens*, in their genus *Nihelia*, the proper generic name for the group must be *Nihelia* Domrow and Baker (1960), not *Hemicheyletus* Lawrence (1961).

Cheletosoma Oudemans

Cheletosoma Oudemans, 1905, Ent. Ber. Nederl. Ent. Ver. 1(21):207.

Females of *Cheletosoma* have no teeth on the palp claw (a nymphal stage and males have 1 tooth). With 1 palptarsal sensillum comblike, 1 plain and 2 sickle-like. Eyes absent. Humeral setae ultralong. Hysterosoma bears a small unpaired dorsal plate situated far posteriorly. Opisthosoma has only 1 pair of ultralong dorsal setae at posterior margin of suranal plate. Tarsi I to IV with claws and empodia.

Type species.—Cheletosoma tyrannus Oudemans, 1905, Ent. Ber. Nederl. Ent. Ver. 1(21):207. Monotypic. From shaft of wing feathers of *Aramus guarauna*, tropical America.

Cheyletiella Canestrini

Cheyletiella Canestrini, 1886, Prospetto dell'Acarofauna Italiana. Vol. II:170.
Cheletiella Oudemans, 1906, Mém. Soc. Zool. France 19:211.
Ewingella Vail and Augustson, 1943, Jour. Parasitol. 29(6):419–421.

Palp claw with many weak teeth. Palp tarsus bears no comblike setae, with simple setae only. Peritremal segments slender in transverse arms, short and wide in descending arms; more than 10 segments. Propodosoma with extensive plate in both sexes. Eyes absent. Dorsal body setae acicular, smooth or finely barbed, never spatulate or fan-shaped. First 3 pairs of dorsolateral propodosomal setae arise on median plate; 4th pair arise on integument close beside humeral setae. With 2 pairs acicular median setae near posterior margin of propodosomal plate. Hysterosoma without a median plate, but may have a pair of rudimentary plates; without setae. Opisthosoma with 1 pair of posterior dorsolaterals smooth and appreciably longer than other dorsal setae. Tarsi I to IV have no paired claws, but each has a conspicuous multirayed empodium.

Type species.—Cheyletus parasitivorax Mégnin, 1878; by monotypy.

Two species are included in *Cheyletiella* at the present time:

Cheyletiella parasitivorax (Mégnin), 1878, Jour. Anat. et. Physiol. 14:425–426. From rabbits, probably France.
Cheyletiella yasguri Smiley, 1965, Proc. Ent. Soc. Wash. 67(2):76–77. From domestic dogs, Ithaca, New York.

Eucheyletiella Volgin

Eucheyletiella Volgin, 1969, Akad. Nauk S.S.S.R., Zool. Inst., Opredel. p. Faune S.S.S.R. no. 101:367–368.

Palp claw without teeth. Palp tarsus with no comblike setae, with simple setae only. Peritremes arch-shaped (greatly reduced in *E. takahasii*, male); peritremal segments not conspicuously widened, segments in transverse arms similar to those of descending arms, with 10 or fewer segments. Propodosoma with a single shield, bearing 3 pairs of barbed lateral setae; 4th pair situated off plate near humeral setae. Propodosomal plate without median setae. Eyes absent. Hysterosoma with dorsal plate, with 3 pairs of long barbed setae and several pairs of short acicular setae near posterior end. Males with a hysterosomal plate. Tarsi I to IV without claws, but with well-developed rayed empodia.

Type species.—Cheyletiella ochotonae Volgin, 1960, Akad. Nauk S.S.S.R., Zool. Inst., Parazitol. Sbornik 19:238–242; by original designation. From *Ochotona macrotis* (Gunth.) (Lagomorpha, Ochotonidae), U.S.S.R.

Two other species are included:

Eucheyletiella takahasii (Sasa and Kano), 1951, Japan. Jour. Expt. Med. 21:205–207. Originally described in *Cheyletiella;* known only from the male and nymph. From *Ochontona hyperborea yesoensis* Kishida, Hokkaido, Japan.
Eucheyletiella johnstoni (Smiley), 1965, Proc. Ent. Soc. Wash. 67(2):77–79. From *Ochotona princeps*, Santa Fe, New Mexico, U.S.A.

According to Volgin (1969), *Eucheyletiella* may be distinguished from *Cheyletiella*, to which it is closely related, by the smaller number and uniform shape of the peritremal segments and the chaetotaxy of the propodosomal shield in females. All three known members of this genus have been taken from lagomorphs in the family Ochotonidae (pikas).

Neocheyletiella Baker

Neocheyletiella Baker, 1949, Proc. U.S. Nat. Mus. 99(3238):271.
Ornithocheyla Lawrence, 1959, Parasitology 49:433–434.

Palp claw inconspicuous, without teeth. Palptarsal sensilla simple, none comblike. Propodosomal shield small, angularly pyriform, restricted to midsection, at most with only 1 pair of median setae on anterolateral margins. Eyes absent. Without hysterosomal shield. Body and appendicular setae acicular or filiform, smooth or with fine, uniform barbs. Most species with 3 or more pairs of ultralong filiform setae. Tarsi compact, with barely developed pedicels. All tarsi with robust claws and multirayed empodia. Coxae II weakly defined or vestigial; coxae IV widely separated from coxae III.

Type species.—Neocheyletiella rohweri Baker, 1949, Proc. U.S. Nat. Mus. 99(3238):274; by original designation. From *Sitta pygmaea melanotis*, Colorado, U.S.A.

Four other species are included:

N. artami Domrow, 1965, Proc. Linn. Soc. N. S. Wales 90:210–212. From dusky wood-swallow, *Artamus cyanopterus* (Latham), Exeter, Tasmania.

N. megaphallos (Lawrence), 1959, Parasitology 49:434–438. From a black-cheeked waxbill, *Estrilda erythronotus*, Bechuanaland, Africa. Originally described as *Ornithocheyla megaphallos* Lawrence; synonymy by Volgin (1966*b*).

N. oudemansi Volgin, 1969, Akad. Nauk S.S.S.R., Zool. Inst., Opredel. p. Faune S.S.S.R. no. 101:417. From the house martin, *Delichon urbica* L., and the bank swallow, *Riparia riparia* L., U.S.S.R.

N. smallwoodae Baker, 1949, Proc. U.S. Nat. Mus. 99(3238):273. From *Leucosticte australis*, Colorado, and the robin, *Turdus migratorius*, Delaware.

All known species have been described from birds. The distinguishing features of the genus are the reduced propodosomal plate, the incomplete development of coxae II, the separation of coxae IV from coxae III, and the dorsomedian gonopore of the male.

Bakericheyla Volgin

Bakericheyla Volgin, 1966*b*, Entomol. Rev. 45(1):121.

Gnathosoma small, with a short, wide rostrum. Palps weakly developed; femur and genu slightly elongated. Palp tarsus poorly developed, with simple setae only. Palp claw without teeth. Peritremes weakly M-shaped, with 5 or 6 elongated segments. Propodosoma with 6 pairs of setae, including humerals; 5 pairs on or adjacent to propodosomal plate. Eyes absent. Hysterosoma without a shield. Body and appendicular setae acicular or filiform, may be finely barbed; dorsum with very short setae interspersed with ultralong setae. Legs short relative to body, legs I and II shorter than legs III and IV. Coxae II distinct, posterior margin may be weakly sclerotized. Coxae IV contiguous with coxae III; coxae IV with 2 pairs of setae. All tarsi with paired claws and multirayed empodium.

Type species.—Cheyletiella chanayi Berlese and Trouessart, 1889, Bull. Biblio. Scient. Ouest 2(9):135; by original designation. From a bird, *Fringilla coelebs*, France; also reported from *Motacilla alba* in Europe.

Volgin and Nikolaeva (1965), Furman and Sousa (1969), and Volgin (1969) add to the host list, redescribe the species, and contribute interesting biological information. According to Furman and Sousa, this species is a blood-sucking ectoparasite. It has the habit of building nests of silk on the skin of its host.

All members of the genus have been described from birds. Six other species are included:

B. faini (Lawrence), 1959, Parasitology 49:431–433. From *Cossypha dichroa*, Cape Province, Africa.

B. heteropalpus (Mégnin). Described as *Cheyletus heteropalpus* by Mégnin, 1878, Jour. Anat. et Physiol. 14:426–427. From pigeons and passerine birds, France (probably).

B. macronycha (Mégnin). Described as *Cheyletus macronychus* by Mégnin, 1878, Jour. Anat. et Physiol. 14:427–428. From passerine birds, Bengal.

B. microrhyncha (Berl. et Trt.). Originally described as *Cheyletiella microrhyncha* by Berlese and Trouessart, 1889, Bull. Biblio. Scient. Ouest. 2(9):121–143. Host not known; from France.

B. subquadrata (Lawrence), 1959, Parasitology 49:427–429. From the little bee-eater, *M. pusillus*, Bechuanaland, Africa.

B. transvaalica (Lawrence), 1959, Parasitology 49:430–431. From the little bee-eater, *Mellitophagus pusillus*, the blue-cheeked bee-eater, *Merops persica*, and *Merops apiaster*, Transvaal, Africa.

Volgin (1966*b*) established *Bakericheyla* and distinguished it from the closely related genus *Neocheyletiella* on the basis of the contiguity of coxae III and IV, and on the numbers of setae on certain segments of legs III and IV: trochanters 2-1, genua 2-2, and tibiae 3-3. The presence of three setae on tibia III and IV is a peculiar character observed in few other genera, and may help to make a discrete separation between *Neocheyletiella* and *Bakericheyla*.

The species *heteropalpus* (Mégnin), *macronycha* (Mégnin), and *microrhyncha* (Berl. et Trt.) now present serious problems in identification because their original descriptions are too general.

Ornithocheyletia Volgin

Ornithocheyletia Volgin, 1964*b*, Zool. Zhurnal 43:28–29.

Closely resembles *Bakericheyla*. Well-developed hysterosomal plate bears 2 to 3 pairs of setae. Coxae IV contiguous with coxae III. Coxae IV with 1 pair setae, trochanters III and IV without setae. Peritremes gently arched, like a bow, with 3 pairs of segments. Dorsal apical setae of tarsi dilated and split distally.

The included species are:

O. dubinini Volgin, 1964*b*, Zool. Zhurnal 43:29–32. Designated as type species.

O. canadensis (Banks), 1909, Proc. Ent. Soc. Wash. 11:133. Redescribed by Volgin.

O. pinguis (Berlese), 1889, Acari..., fasc. 56, no. 3.

KEY TO FEMALES OF ORNITHOCHEYLETIA
(Adapted from Volgin, 1964b)

1. Posterior end of idiosoma with 1 pair of long, smooth setae; tarsus I without humplike swelling ....*pinguis* (Berlese)
   Posterior end of idiosoma with 2 pairs of long, smooth setae (*Thysanura*-like); tarsus I with humplike swelling 2
2. With 4 pairs of setae on propodosomal plate; 3 pairs on hysterosomal plate ....*dubinini* Volgin
   With 1 pair of setae on propodosomal plate; 2 pairs on hysterosomal plate

canadensis (Banks)

Criokeron Volgin

Criokeron Volgin, 1966*b*, Entomol. Rev. 45(1):119.

This genus was created to separate the peculiar *quinta* from *Nihelia* species having no hooklike processes on the gnathobase (basis capituli). The recognitional characters of *Criokeron quinta* (Domrow and Baker) occur on the gnathosoma. The descending limits of the peritremes are exceedingly broad and have many chambers or crosswalls. A pair of large hooklike processes project downward from the gnathosoma. These processes have numerous annulations or cross-set striae.

Type species.—Nihelia quinta Domrow and Baker, 1963; by original designation. From a tree-shrew, *Tupaia glis,* Malaya.

Chelacheles Baker

Chelacheles Baker, 1958, Proc. Ent. Soc. Wash. 60(5):234–235.

Elongate, slender-bodied forms. Gnathosoma relatively small in relation to body size. Palp claw short, nearly straight, with 3 or 4 basal denticles. Palp tarsus bears 2 sickles, 1 heavy outer comb and 1 inner seta which may be comblike or plain. Idiosoma markedly constricted at humeral sulcus and without obvious dorsal plates. Eyes present. Coxae I and II widely separated from coxae III and IV. Opisthosoma vestigial, scarcely more than an anogenital protuberance appended to metapodosoma. Body setae acicular or flagelliform, smooth, several pairs ultralong. Tarsi I to IV provided with paired claws and rayed empodia.

Type species.—Chelacheles strabismus Baker, 1958, Proc. Ent. Soc. Wash. 60(5):235; by original designation.

KEY TO FEMALES OF CHELACHELES

1. Humeral setae not emphatically longer than adjacent dorsal setae of propodosoma
   strabismus Baker
   Humeral setae ultralong, more than 3 times as long as adjacent dorsal setae of propodosoma 2
2. Humeral setae are only ultralong setae on upper part of body proper ....*michalskii* Samšiňák
   Humeral setae of propodosoma and 1st pair of dorsolateral setae of hysterosoma are ultralong ....*bipanus,* n. sp.

Chelacheles bipanus, new species
(Fig. 58)

Palp claw pointed, shorter than tibial joint, 3 or 4 basal teeth. Palp tarsus sessile; a minute pointed spine projects from its dorsal rim. Outer comb ca. 10 teeth; sensillum corresponding to inner comb possibly has several exceedingly fine barbs on its convex curvature. Palp femur with 1 flagelliform dorsal seta (45). Stylophore almost triangular in outline, surface with no obvious ornamentation; protegmen subconic, its pointed apex fitted into gutter of rostrum. Peritremes obtusely bent where short transverse arms give rise to much longer descending arms, 7 to 9 links per side. Basis capituli retains only a vestige of midventral keel-like apodeme so prominent in other cheyletids; this remnant is a short bar projecting forward from ventral rim of basis. Body integument pliable, longitudinally striated except for a wide girdle of transverse striae in mid-section. Incipient propodosomal shield represented by a faintly tanned patch of cuticle shaped like inverted triangle, its base line wedged between eyes, apex directed rearward between 3d pair of setae. Dorsal body setae acicular, apparently smooth, 13 pairs plus humerals; humerals and 1st marginals on hysterosoma flagelliform, subequal, ultralong (ca. 90); all other dorsals comparatively short (12–23). Paragenitals 1 pair, genitals 2 pairs, anals 3 pairs.

Length ratio: leg I/idiosoma = 0.4. Setae on legs I to IV: coxae 2-1-2-1, femora 2-2-1-1, genua 3-2-2-2, tibiae 5-4-3-3, tarsi 8-7-6-6. Dorsal seta on femora I to IV ultralong (55–70). Guard seta not discernible on tarsus I. Measurements (n = 4): length idiosoma 369, gnathosoma 70, leg I 134, tarsus I 43.

Holotype.—Female, from willow bark and twigs, American Canyon, Solano Co., Calif., Feb. 15, 1951 (S. F. Bailey); 3 female paratypes, 1 male, same collection. One paratype female in USNM; other types in UCD.

The type species, *strabismus* Baker, from Portugal, and this species collected in California are very similar in most respects. The two species are distinguishable according to differences in the lengths of certain setae: in *bipanus* the humeral, the first lateral hysterosomal, and the dorsal seta on each of the femora are emphatically longer than the corresponding setae of *strabismus*.

The two previously described species are:

Chelacheles strabismus Baker, 1958, Proc. Ent. Soc. Wash. 60(5):235. Collected in flour mill, Lisbon, Portugal.

Chelacheles michalskii Samšiňák, 1962, Čas. Čs. Spol. Ent. (Acta Soc. Ent. Českoslov.) 59(2):183–185. Taken from galleries of bark beetles, *Scolytus multistriatus* Marsham and *S. pygmaeus* F., beneath bark of elm, *Ulmus campestris* L., Prague, Czechoslovakia.

Bak Yunker

Bak Yunker, 1961, Canadian Ent. 93(11):1023.

Slender-bodied, almost cylindrical forms in which legs I and II are very distant from legs III and IV. Opisthosoma diminutive, scarcely more than an end-cap appended to 4th leg-bearing segment of metapodosoma. Legs IV directed rearward, with anogenital apparatus flanked by their enlarged coxal segments. Gnathosoma robust, segments compact, articulated segments of pedipalps wider than long; 2 comblike and 2 acicular seta on palp tarsus; tibial claw and strong tarsal sensilla contribute almost half the length of this gnathal appendage. Peritremes loop far forward before sharply bending rearward. Dorsal cuticula with feeble sclerotization comprising a faintly tanned propodosomal shield of considerable size and several small ill-defined platelets on metapodosoma. No eyes. Ambulatory legs stubby; coxae III and IV coalesced, forming an integral sclerite on each side. Tarsi I to IV bear claws and rayed empodia.

Type species.—Bak sanctaehelenae Yunker, 1961; by original designation.

KEY TO FEMALES OF *BAK*

1. Each peritreme has a median descending arm comprising 3 links; claws on legs I to IV sharply bent, each with a basal swelling ....*sanctaehelenae* Yunker
   Each peritreme bends forward from point of origin; claws on legs I to IV smooth, C-shaped hooks 2
2. Tibiae 5-4-4-4; genu III with 1 seta .... *deleoni* Yunker
   Tibiae 4-3-3-3; genu III without setae ....*micidus*, n. sp.

Bak sanctaehelenae Yunker

(Fig. 59*b–d*)

Bak sanctaehelenae Yunker, 1961, Canadian Ent. 93(11):1023–1028.

Palp claw a robust crescentic spur, slightly bent at base and close to blunt tip, basal part with a broadly rounded boss and one finger-like tooth. Outer comb longer than claw, ca. 20 teeth plus extended blade; inner comb ca. 28 teeth. Stylophore covers rostrum almost to nipples of superior adoral setae, faintly striate lengthwise, with a few minute pores (ca. 40) dispersed over surfaces of tegmen and protegmen. Protegmen flanked by ascending arms of peritremes; its anteriormost lip a transverse fold supported between opposite cornuae of peritremes. Peritremes have two abrupt bends; each peritreme has a median descending arm (3 links) directed backward from origin; it then reverses direction and its ascending arm (5 links) extends diagonally forward to an acute bend (cornua) from which a lateral descending arm (6 links) courses backward. Dorsum usually without plating; a feebly sclerotized propodosomal plate covers tergal area bounded by 4 pairs of propodosomal setae; 3 small, barely discernible plates occur on metapodosoma between legs III, and vestiges of 3 median platelets bear 3 pairs of opisthosomal setae. Dorsal setae spiniform, very short, 11 pairs plus humerals; humerals ultralong (ca. 85), flagelliform. Ultralong flagelliform setae on coxae I, III, and IV; 3 pairs of setae on ventral integument, middle pair ultralong (ca. 70). Paragenitals 2 pairs, genitals 2 pairs, anals 1 pair only.

Length ratio: leg I/idiosoma = 0.4. Setae on legs I to IV: coxae 2-1-(2-1), femora 2-2-2-1, genua 2-1-1-0, tibiae 5-4-4-4, tarsi 10-8-7-7. Tarsus I: solenidion *w*I diminutive (5); guard seta flagelliform, fairly long (35), originating a short distance behind *w*I. Pretarsi I to IV with claws robust and sharply bent into compressed hooks; each claw with a basal protuberance. Measurements (n = 1): length idiosoma 413, gnathosoma 121, leg I 172, tarsus I 74.

Collection data.—California: one specimen from topsoil near Cuyamaca Lake, San Diego Co. (L. M. Smith); holotype female and paratypes collected on Mt. St. Helena, Napa Co., in dense woods (S. F. Bailey).

The shape of the peritremes is believed to be a useful spot character for this species.

The peculiar anantomy of the *Bak* species evokes speculative thought about its adaptive significance. Quite possibly these species are molelike burrowing forms. The strong anteriorly directed palp claw, the cylindrical body, the unusual carriage of the legs with their rugged claws, and the absence of ornate body setae are indicative features. Furthermore, the gnathal appendages appear to be especially fitted for stresses of forward motion against resistance. The palp segments are massive (i.e., muscular) but short, the rostrum is a stubby "beak," and the protegmenal part of the stylophore appears to have thrust backward to reshape the tegmen, and thus distort the lateral sweep of the peritremes characteristic of other cheyletids.

Bak micidus, new species

(Fig. 59*a*)

Palp claws, chaetotaxy, and general organization of gnathosoma substantially as described for *sanctaehelenae*. Peritremes turn forward from origin and ascend to sharp-angled cornuae, possibly 5 links in each ascending arm; descending arms follow marginal contour of tegmen, with 6 or 7 slender links. Stylophore faintly striate lengthwise, with a very few widely dispersed pores (ca 12); front of protegmen appears as a very delicate transverse line or fold of membrane connecting cornuae of each side. Propodosomal plate indistinctly outlined, probably not wide enough to encompass 3d and 4th pairs of dorsolateral setae; a pair of small setiferous plates and 1 median plate on dorsum of metapodosoma in line with legs III. Dorsal setae very short, 11 pairs plus humerals; individual setae of 3 hindmost pairs (opisthosomals) very close together. Humeral setae in pleuroventral position, flagelliform, relatively long (at least 55); midventrals about as long as humerals. Paragenitals 1 pair, genitals 1 pair, anals 2 pairs.

Length ratio: leg I/idiosoma = 0.4. Setae on legs I to IV: coxae 2-1-(1-1), femora 1-1-1-1, genua 2-0-0-0, tibiae 4-3-3-3, tarsi 9-8-7-7. Tarsus I: mesal and lateral addorsals *tc* about same length, but mesal addorsal about twice as thick, blunt-tipped; lateral addorsal slender, flagelliform. All tarsi bear smooth, crescentic or C-shaped hooklets, without basal protuberances. Measurements (n = 1): length idiosoma 327, gnathosoma 101, leg I 125, tarsus I 51.

Holotype.—Female, from rotted leaves of *Tourneyia*, Old Bella Vista Trail, Santa Cruz, Galapagos Islands, Feb. 12, 1964 (R. O. Schuster); deposited in UCD.

This species closely resembles *deleoni* Yunker. In *micidus* the two addorsal setae *tc* are dissimilar; genu III lacks setae; tibiae I to IV bear 4-3-3-3 setae, respectively; and the anal cover bears two pairs of setae. In *deleoni* the addorsals on tarsus I are similar to each other; genu III has one seta; tibiae I to IV are provided with 5-4-4-4 setae, respectively; and the anal cover has but one pair of setae.

Yunker's other species of *Bak* is:

Bak deleoni Yunker, 1961, Canadian Ent. 93(11):1028–1031. On *Casuarina*, Coral Gables, Florida.

ACKNOWLEDGMENTS

If this paper helps in the development of a workable taxonomy for the mite family Cheyletidae, some credit may be given to the many interested persons who have assisted and encouraged the authors in its preparation. The senior author was accorded guest privileges in the laboratories of: Dr. Edward W. Baker, Division of Insect Identification and Parasite Introduction, Agricultural Research Service, United States Department of Agriculture, Washington, D.C.; Dr. Martin H. Muma, Department of Entomology, Citrus Experiment Station, University of Florida, Lake Alfred, Florida; and Dr. Harold A. Denmark, Chief of the Entomology Section, Division of Plant Industry, Florida Department of Agriculture, Gainesville, Florida.

Valuable type specimens were loaned to us by Dr. H. W. Levi, Curator in Arachnology, Museum of Comparative Zoology, Harvard University, Cambridge, Massachusetts, and Dr. David C. Lee, Curator of Arachnology, South Australian Museum, Adelaide, South Australia.

We are most grateful to all these gentlemen and to their respective institutions for their generosity in providing research facilities, as well as valuable specimens and stimulating ideas.

LITERATURE CITED

ANDRÉ, M.
1933. Notes sur les Acariens observés dans les magasins regionaux de Tabacs. Ann. des Ephiphytes, Paris, 19(6):331–356.

BAKER, E. W.
1949. A review of the mites of the family Cheyletidae in the United States National Museum. Proc. U.S. Nat. Mus. 99 (3238):267–320.
1958. *Chelacheles strabismus*, a new genus and species of mite from Portugal. (Acarina, Cheyletidae). Proc. Ent. Soc. Wash. 60(5):234–235.

BAKER, E. W., and G. W. WHARTON
1952. An Introduction to Acarology. Macmillan, N.Y. 465 pp.

BANKS, NATHAN
1902. New genera and species of acarians. Canadian Ent. 34:171–176.
1904. A treatise on the Acarina, or Mites. Proc. U.S. Nat. Mus. 28(1382):1–114.
1906. Descriptions of some new mites. Proc. Ent. Soc. Wash. 7:133–142.
1909. New Canadian mites. *Ibid.*, 11:133–144.

BEER, R. E., and D. T. DAILEY
1956. Biological and systematic studies of two species of Cheyletid mites, with a description of a new species (Acarina, Cheyletidae). Univ. Kansas Sci. Bull. 38, part 1(5):393–437.

BERLESE, A.
1882–1893. Acari, Myriapoda et Scorpiones hucusque in Italia reperta, Ordo Prostigmata. Patavii.
1913. Acari nuovi. Manipoli VII–VIII. Redia 9:77–111.
1921. Centuria quinta di Acari nuovi. Redia 14:143–195.

BERLESE, A., and E. L. TROUESSART
1889. Diagnoses d'acariens nouveaux ou peu connus. Bull. Biblio. Scient. Ouest. II, 2(9):121–143.

CANESTRINI, G.
1885. Prospetto dell'Acarofauna Italiana. Vol. II:159–311. Padova.

CANESTRINI, G., and F. FANZAGO
1876. Nouvi acari italiani (ser. 2). Atti Soc. Veneto-Trentina Sci. Nat. 5(fasc. 1):99–111.

COCKERELL, T. D. A.
1917. Arthropods in Burmese amber. Psyche 24:40–45.

COOPER, K. W.
1946. The occurrence of the mite *Cheyletiella parasitivorax* (Mégnin) in North America, with notes on its synonymy and "parasitic" habit. Jour. Parasitol. 32(5):480–482.

CUNLIFFE, F.
1955. A proposed classification of the Trombidiforme mites (Acarina). Proc. Ent. Soc. Wash. 57(5):209–218.
1962. New species of Cheyletidae (Acarina). *Ibid.*, 64(3):197–202.

DE GEER, C.
1778. Mémories pour servir à l'histoire des Insectes. Vol. 7. Stockholm.

DE LEON, DONALD
1962. Three new genera and seven new species of Cheyletids (Acarina: Cheyletidae). Florida Entomologist 45(3):129–137.
1967. Some Mites of the Caribbean Area. Lawrence, Kansas: Allen Press, Inc. 60 pp.

DOMROW, R.
1960. The genus *Chelonotus* Berlese (Acarina, Cheyletidae). Acarologia 2(4):456–460.
1965. Some mite parasites of Australian birds. Proc. Linn. Soc. N. S. Wales 90(2):190–217.

DOMROW, R., and E. W. BAKER
1960. Malaysian parasites. XLIX. A new genus of mites from a Thai mongoose (Acarina, Cheyletidae). Stud. Inst. Med. Res. Malaya 29:194–197.
1963. The genus *Nihelia* (Acarina, Cheyletidae). Acarologia 5(2):225–231.

DUBININ, V. B.
 1957. New classification of mites of the superfamilies Cheyletoidea W. Dub. and Demodicoidea W. Dub. (Acariformes, Trombidiformes). Akad. Nauk S.S.S.R., Zool. Inst., Parazitol. Sbornik, 17:71–136.

EHARA, SHÔZÔ
 1962. Mites of greenhouse plants in Hokkaido, with a new species of Cheyletidae. Annotationes Zool. Japonenses 35(2):106–111.

EWING, H. E.
 1909. North American Acarina. Trans. St. Louis Acad. Sci., 18(5):53–77.

FURMAN, D. P., and O. E. SOUSA
 1969. Morphology and biology of a nest-producing mite, *Bakericheyla chanayi* (Acarina: Cheyletidae). Ann. Ent. Soc. Amer. 62(4):858–863.

GERSON, URI
 1967. Some cheyletid and pseudocheylid mites from Israel. Acarologia 9(2):359–369.
 1968. *Caudacheles*, a new genus in the family Cheyletidae (Acarina: Prostigmata). *Ibid.*, 10(4):645–649.

HAARLOV, N.
 1942. A morphologic, systematic, ecological investigation of Acarina and other representatives of the microfauna of the soil around Mørkelford, Northeast Greenland. Medd. Grønland, no. 1281:171.

HALLER, G.
 1884. Beschreibung einiger neuen Milben. Arch. f. Naturgesch. 50(2):217–235.

HARA, JUN
 1955. A new species of cheyletid mite found on tatami, Japanese rice straw matting. Japan. Jour. Expt. Med. 25:69–70.

HARA, JUN, and MINORU HANADA
 1960. On a newly recorded mite, *Eutogenes narashinoensis* new species, from Japan (Acarina: Cheyletidae). Japan. Jour. Sanitary Zool. 11(1):25–27.

HELLER, A.
 1880. Die Schmarotzer mit besonderer Berücksichtigung der für den Menschen wichtigen. Die Naturkrafte 30:186–188.

HESSLING, THEODORE VON
 1852. Einige notizen über den Weichselzopf. Illustr. Med. Zeit. 1(5):255–259.

HUGHES, A. M.
 1948. The mites associated with stored food products. Ministry of Agriculture and Fisheries. London. 168 pp.
 1961. The mites of stored food. Ministry of Agriculture, Fisheries and Food Tech. Bull. 9. London. 287 pp.

HUGHES, T. E.
 1958. The respiratory system of the mite *Cheyletus eruditus* (Schrank, 1781). Proc. Zool. Soc. London, 130(part 2):231–239.
 1959. Mites, or the Acari. Univ. of London, The Athlone Press. 225 pp.

KARPELLES, L.
 1884. Beiträge zur Naturgeschichte der Milben. Berl. Entom. Zeitschr. 28:1–34.
 1884. Neue Milben. *Ibid.*, 28(2):231–244.

KOCH, C. L.
 1839. Deutschlands Crustaceen, Myriapoden und Arachniden. Fasc. 23. Regensberg.

LABOULBENE, A.
 1851. Description de quelques Acarines et d'une Hydrachne. Ann. Soc. Ent. France (ser. 2) 9:302.

LATREILLE, A.
 1796. Précis de caractères générique des insectes, disposés dans un ordre naturel. An. 5 (1797). Paris.

LAWRENCE, R. F.

1948. Some new pilicolous mites from South African mammals. Parasitology 39:39–42.

1954. The known African species of Cheyletidae and Pseudocheylidae (Acarina, Prostigmata). Ann. Natal Mus. 13(1):65–77.

1959. New mite parasites of African birds (Myobiidae, Cheyletidae). Parasitology 49(¾): 416–438.

1961. In F. Zumpt, ed. The arthropod parasites of vertebrates in Africa south of the Sahara. Publ. South African Inst. Med. Res. 11:1–457.

LEACH, W. E.

1815. A tabular view of the external characters of four classes of animals, etc. Trans. Linn. Soc. London, 11:399.

McGREGOR, E. A.

1956. The mites of citrus trees in southern California. Mem. So. Calif. Acad. Sci. 3(3):1–24.

MÉGNIN, P.

1878. Mémoire sur un nouveau groupe d'Acariens, les Cheylétides parasites. Jour. de l'Anat. et Physiol. (Paris) 14:416–441.

MICHAEL, A. D.

1878. On a new species of Acarus of the Genus *Cheyletus*, believed to be new. Jour. Roy. Micr. Soc. 1:133–138.

MOLA, PASQUALE

1907. Nuovi acari parasiti. Zool. Anz. 32:41–44.

MOQUIN-TANDON, H. B. A.

1863. Eléments de Zool. médicale, p. 314.

MUMA, M. H.

1964. Cheyletidae (Acarina: Trombidiformes) associated with Citrus in Florida. Florida Entomologist 47(4):239–253.

NEWSTEAD, R., and H. M. DUVALL

1918. Bionomic, morphological and economic reports on the acarids of stored grain and flour. Rep. Grain Pests Comm., Roy. Soc. London, no. 2, pp. 1–52.

NORDENSKIÖLD, ERIK

1900. Anteckningar om Acarider samlade i hö. Medd. Soc. Faun. Fenn. 26:34–38.

OUDEMANS, A. C.

1897. List of Dutch Acari. IV. Tijdschr. Ent. 40:111–135.

1903a. Tijdschr. Nederl. Dierk. Ver. (ser. 2) 8:16.

1903b. Acarologische Aanteekeningen. VI. Ent. Ber. Nederl. Ent. Ver. 1(12):84.

1903c. Notes on Acari. XI. Tijdschr. Ent. 46:93–134.

1904a. Acarologische Aanteekeningen. XI. Ent. Ber. Nederl. Ent. Ver. 1(17):154

1904b. Acarologische Aanteekeningen. XII. *Ibid.*, 1(18):160–163.

1904c. Notes on Acari. Tijdschr. Nederl. Dierk. Ver. (ser. 2) 8:xv–xvi.

1904d. Acarologische Aanteekeningen. XIII. Ent. Ber. Nederl. Ent. Ver. 1(19):169–174.

1905. Acarologische Aanteekeningen. XV. *Ibid.*, 1(21):207–210.

1906. Révision des Chélétinés. Mém. Soc. Zool. France 19:36–218.

1914. Acarologische Aanteekeningen. LIV. Ent. Ber. Nederl. Ent. Ver. 4 (78):101–103.

1928. Acarologische Aanteekeningen. XCIII. *Ibid.*, 7(162):343.

1937. Kritisch historisch Overzicht der Acarologie. Derde Gedeelte, Band C: 799–1348. Leiden: E. J. Brill.

1938. Notes on Acari. Tijdschr. Ent. 81: Versl. lxxv.

PACKARD, A. S.

1878. A Guide to the Study of Insects. New York: Henry Holt and Co. 715 pp.

PELAEZ, D.

1962. *Hoffmannita mexicana* gen. et sp. nov., Cheyletidae parasito de un alacrán del género *Centruroides* (Acarina, Prostigmata). Anales de la Escuela Nacional de Ciencias Biologicas 11:71–78.

POPPE, S. A.

1888. Uber parasitische milben. Abh. naturwiss. Verein, Bremen 10:205–240.

RADFORD, C. D.
1950. Systematic check list of mite genera and type species. Int. Union Biol. Sci. (ser. C) no. 1: 1–232.

ROHDENDORF, B. B.
1940. Mites of the families Cheyletidae and Pediculoididae (in Russian). Moscow Univ. Uchenye zapiski, Wissensch. Ber. 42:69–98.

SAMŠIŇÁK, K.
1956. Die Milben (Acari) auf der Fliege *Laphria flava* L. Mem. Soc. Zool. tchécosl. 20:353–357.
1959. Contribution à l'étude des Acariens de Madagascar. Mém. Inst. Sci. Madagascar 13A: 81–85.
1962. *Chelacheles michalskii* n. sp. (Acari, Cheyletidae). Časopis Československé Společnosti Entolomogické (Acta Soc. Ent. Československ.) 59(2):183–185.

SASA, MANABU, AND ROKURO KANO
1951. *Cheyletiella takahasii*, n. sp., a new species of parasitic mite from *Ochotona* of Hokkaido, Japan (Acarina, Cheyletidae). Japan. Jour. Expt. Med. 21:205–207.

SCHRANK, F. P.
1781. Enumeratio insectorum Austriae indigenorum. Aug. Vindel.

SHAW, G.
1794. Vivarium Naturae, or the Naturalist's Miscellany. Vol. 6, pl. 187.

SMILEY, R. L.
1965. Two new species of the genus *Cheyletiella* (Acarina: Cheyletidae). Proc. Ent. Soc. Wash. 67(2):75–79.

SMILEY, R. L., AND J. C. MOSER
1970. Three cheyletids found with pine bark beetles (Cheyletidae: Acarina). Proc. Ent. Soc. Wash. 72:(2):229–236.

SNODGRASS, R. E.
1948. The feeding organs of Arachnida, including mites and ticks. Smithson. Misc. Coll. 110 (10). 93 pp.

SOUTHCOTT, R. V.
1961. Studies on the systematics and biology of the Erythraeoidea (Acarina), with a critical revision of the genera and subfamilies. Australian Jour. Zoology 9(3):367–610.

SUMMERS, F. M., AND E. I. SCHLINGER
1955. Mites of the family Caligonellidae (Acarina). Hilgardia 34(12):539–561.

THEWKE, SIEGFRIED, AND W. R. ENNS
1968. A new species of predaceous mite (Acarina: Cheyletidae) from galleries of bark beetles in Missouri. Acarologia 10(2):215–219.

TROUESSART, E. L.
1885. Description d'un nouveau genre et d'une nouvelle espèce de la sous-familie des cheyletiens. Bull. Soc. Étude Sci. d'Anger 14:90–91.

VAIL, E. L., AND G. F. AUGUSTSON
1943. A new ectoparasite (Acarina: Cheyletidae) from domestic rabbits. Jour. Parasitol. 29(6):419–421.

VITZTHUM, H.
1920. Acarologische Beobachtungen (Zweite Reihe). Arch. f. Naturgesch. 84A(6):1–40.
1931. Acari. *In* Handbuch der Zoologie. Ed. by W. Kükenthal and T. Krumbach. Band 3, Hälfte 2, Lieferung 1:1–160.

VOLGIN, V. I.
1949. Materials for the systematics of predaceous mites of the genus *Cheyletus* Latr. C. R. Acad. Sci. U.S.S.R. Moscow, N. S. 64:583–586.
1955. Acarina of the rodent fauna of the U.S.S.R. Family Cheyletidae. Akad. Nauk S.S.S.R. Zool. Inst., Opredel. p. Faune S.S.S.R. no. 59:152–176.
1958. On the taxonomy of predatory mites of the family Cheyletidae. I. Genus *Eutogenes* Baker, 1949. Entomol. Obozrenie 37(2):460–463.

1960. On the taxonomy of predatory mites of the family Cheyletidae. II. *Cheyletiella*. Akad. Nauk. S.S.S.R., Zool. Inst., Parazitol. Sbornik 19:237–248.

1961. On the taxonomy of predatory mites of the family Cheyletidae. III. Genus *Cheletacarus* Volgin, gen. nov. *Ibid.*, 20:248–256.

1962. On the taxonomy of predatory mites of the family Cheyletidae. IV. Genus *Neoacaropsis* Volgin, gen. nov. Entomol. Rev. 41(3):425–428.

1963a. On the taxonomy of predatory mites of the family Cheyletidae. IV [V]. Genus *Euchey-letia* Baker, 1949. Akad. Nauk S.S.S.R., Zool. Inst., Parazitol. Sbornik 21:44–68.

1963b. Two new genera of predatory mites of the family Cheyletidae (Trombidiformes). Entomol. Rev. 42(4):504–508.

1964a. On the taxonomy of predatory mites of the family Cheyletidae VII. Genus *Neoeucheyla* Radford, 1950. Akad. Nauk S.S.S.R., Zool. Inst., Parazitol. Sbornik 22:88–97.

1964b. On the taxonomy of predatory mites of the family Cheyletidae. VI. Genus *Ornitho-cheyletia* Volgin, gen. nov. Zool. Zhurnal 43(1):28–36.

1965. General characteristics of the genus *Cheletophyes* Oudms. (Acarina, Cheyletidae). Akad. Nauk S.S.S.R., Trudy Zool. Inst. 35:288–299.

1966a. A new genus and species of mites of the family Cheyletidae Leach (Acarina, Trombidiformes). *Ibid.*, 37:277–285.

1966b. Morphological peculiarities of cheyletid mites (Acarina, Trombidiformes) and their ontological development. Entomol. Rev., 45(1):118–124.

1969. Acarina of the family Cheyletidae, world fauna. Akad. Nauk S.S.S.R., Zool. Inst., Opredel. p. Faune S.S.S.R. no. 101:1–432.

VOLGIN, V. I., and I. I. NIKOLAEVA

1965. On parasitism of predaceous mites of the genus *Neocheyletiella* Baker, 1949 (Acarina, Cheyletidae). Akad. Nauk S.S.S.R., Trudy Zool. Inst. 35:300–304.

WAFA, A. K. and Z. R. SOLIMAN

1968. Five genera of family Cheyletidae (Acarina) in the U.A.R. with a description of four new species. Acarologia 10(2):220–229.

WOMERSLEY, H.

1941. Notes on the Cheyletidae (Acarina, Trombidoidea) of Australia and New Zealand, with descriptions of a new species. Rec. So. Australian Mus. 7(1):51–64.

1955. A new genus and two new species of Acarina from Northern Australia. Proc. Linn. Soc. N. S. Wales 80(3):214–216.

YUNKER, C. E.

1960. *Alliea laruei* n. gen., n. sp. (Acarina: Cheyletidae) from *Rattus norvegicus* (Erxleben) in Florida. Proc. Helminth. Soc. Wash. 27(3):278–281.

1961. The genera *Bak*, new genus, and *Cheletomimus* Oudemans, with descriptions of three new species (Acarina: Cheyletidae). Canadian Ent. 93(11):1023–1035.

ZACHVATKIN, A. A.

1935. A Short Key to the Granary Mites, pp. 1–31. 2d ed. Moscow (in Russian).

ZAHER, M. A., and I. R. SOLIMAN

1965. *Eutogenes punctata* n. sp. (Acarina: Cheyletidae). Bull. Soc. Ent. Egypte 49:65–66.

1967. The family Cheyletidae (Acarina) in the U.A.R., with a description of four new species. *Ibid.* 51:21–26.

FIGURES 1–59

Scale of Drawings

The magnification of many of the illustrations is indicated by a linear scale drawn close beside the figure to which it applies. Each scale represents 100μ.

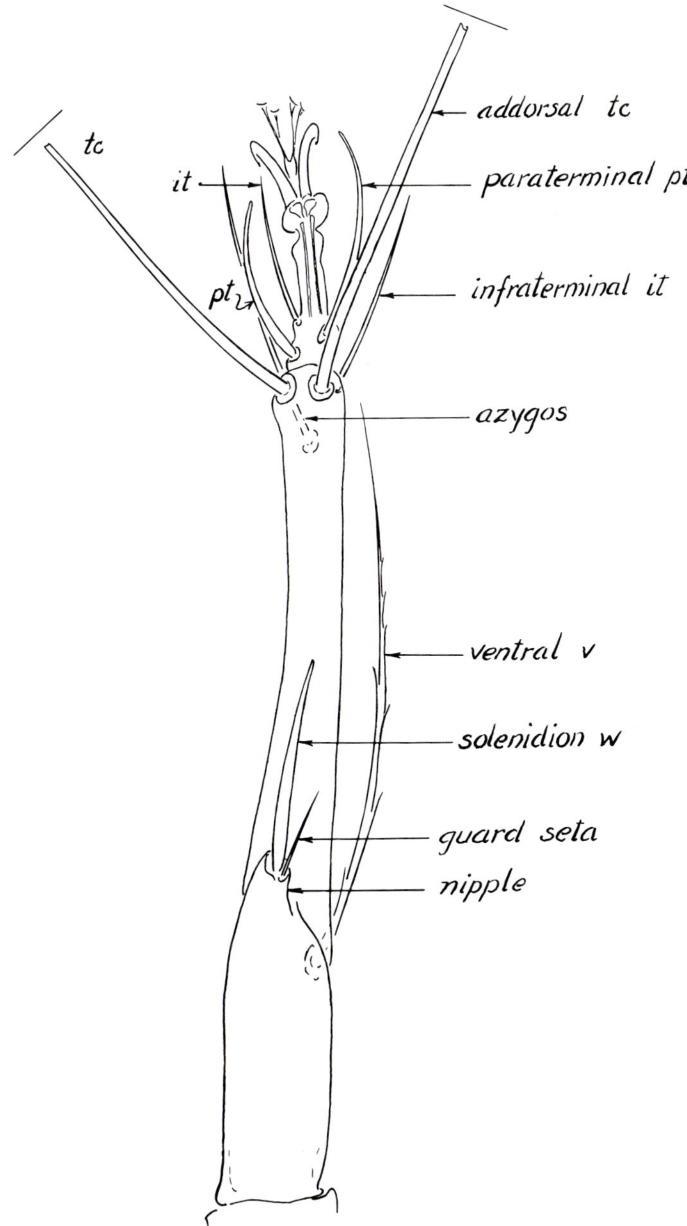

Fig. 1. Diagram of tarsus ɪ with 10 setiform sensilla, as in *Cheyletus*. The 4 subterminal setae adjoining pretarsus (claws and empodium) arise on a tapered part of tarsus referred to as *pedicel*. Paired condition of *parater minal* setae *pt* (upper lateral) and *infraterminal* setae *it* (below) often obscure because these sensilla are not always side by side. Azygos seta appears to be transient, identifiable in some genera, not in others.

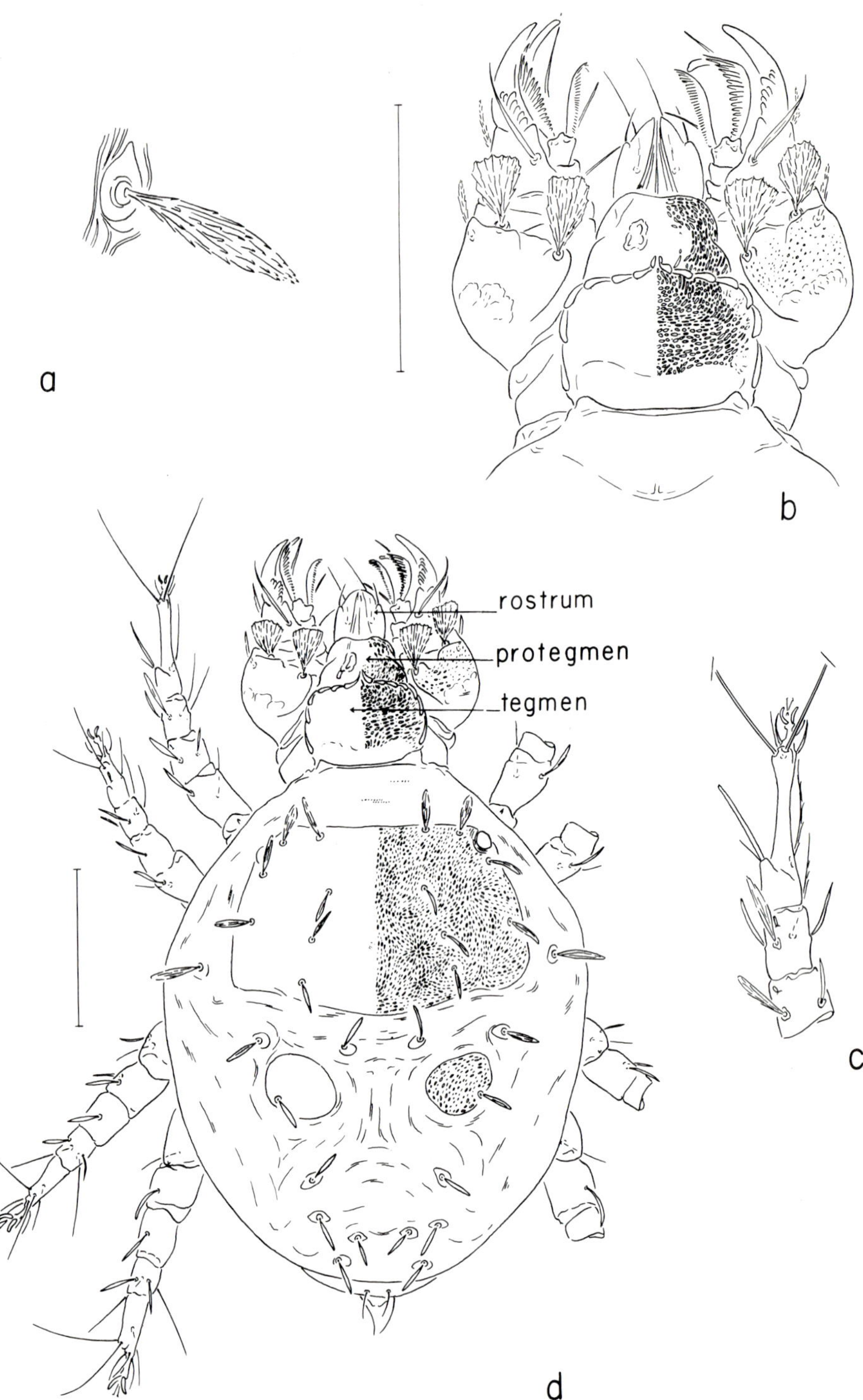

Fig. 2. *Cheletomimus berlesei* (Oudemans). *a*, Humeral seta; *b*, gnathosoma; *c*, distal segments of left leg I; *d*, dorsal aspect of female.

Fig. 3. *Cheletomimus duosetosus* Muma. *a*, First dorsolateral hysterosomal seta; *b*, gnathosoma; *c*, left leg I; *d*, dorsal aspect of female.

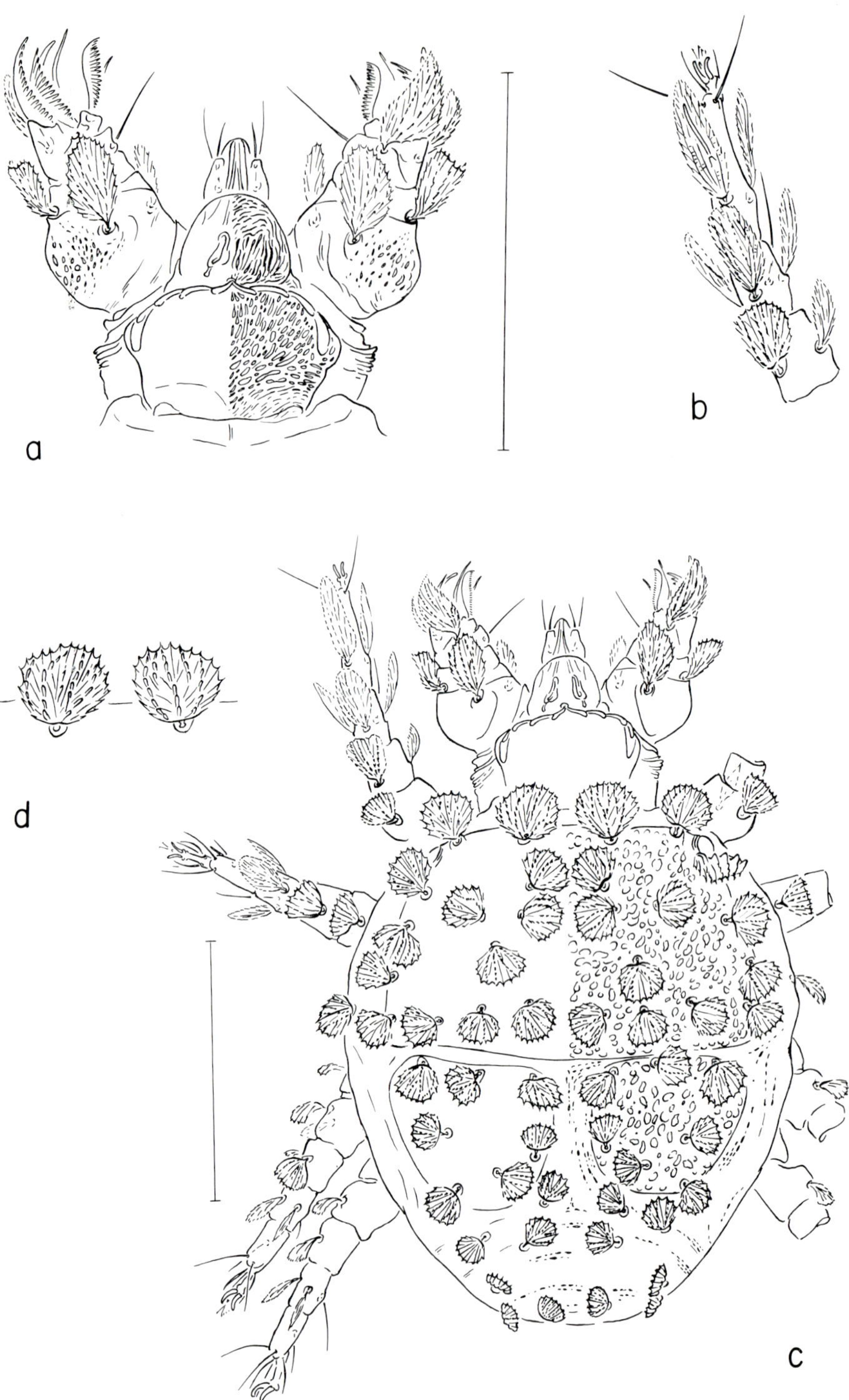

Fig. 4. *Oudemansicheyla denmarki* (Yunker). *a*, Gnathosoma; *b*, distal segments of left leg I; *c*, dorsal aspect of female; *d*, first pair of propodosomal setae (i.e., verticals).

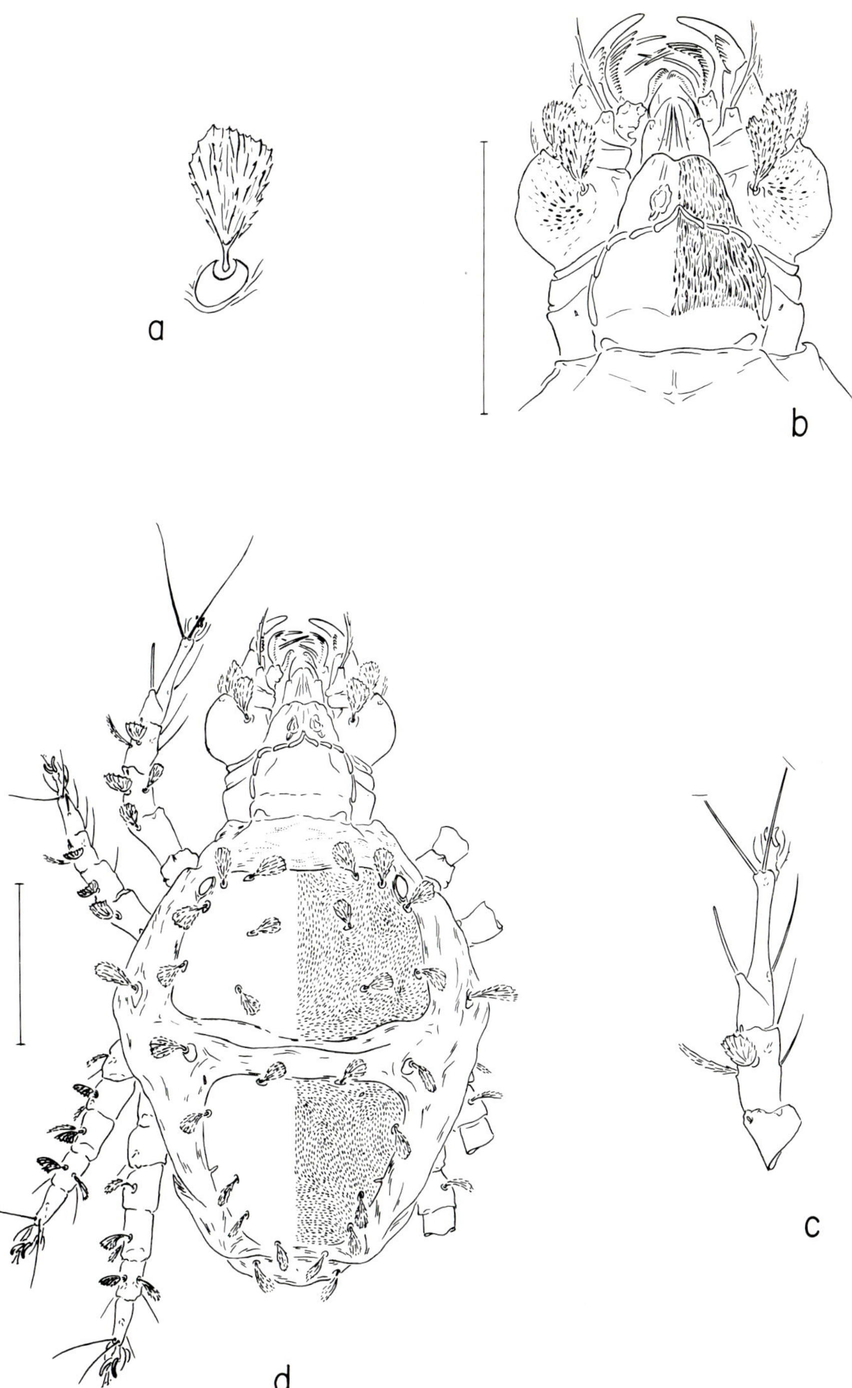

Fig. 5. *Hemicheyletia bakeri* (Ehara). *a,* First dorsolateral hysterosomal seta; *b,* gnathosoma; *c,* left leg I; *d,* dorsal aspect of female.

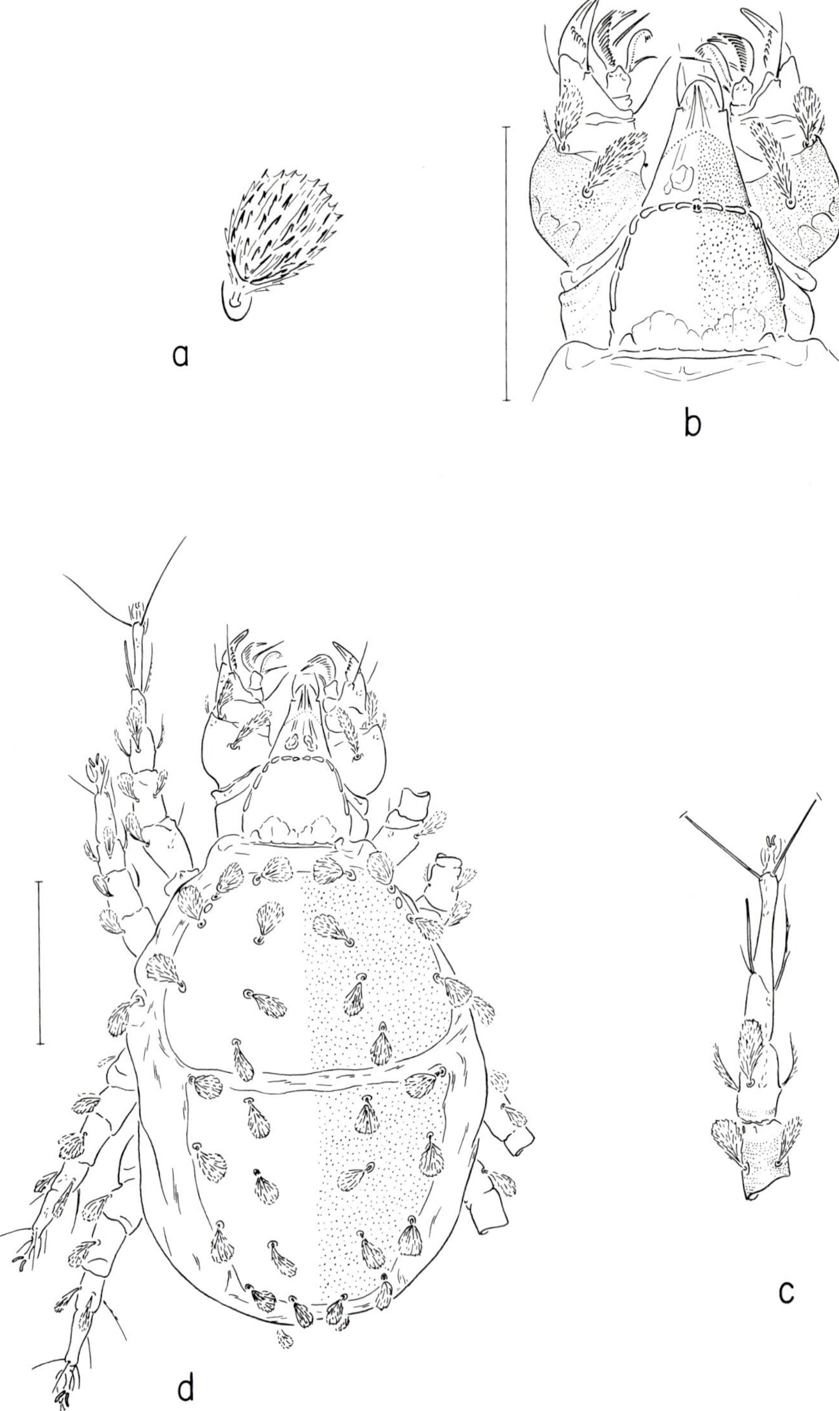

Fig. 6. *Hemicheyletia granula*, n. sp. *a*, First dorsolateral hysterosomal seta; *b*, gnathosoma; *c*, left leg I; *d*, dorsal aspect of female.

Fig. 7. *a, Hemicheyletia congensis* (Cunliffe); *b, Hemicheyletia volgini* (Cunliffe).

Fig. 8. *Hemicheyletia cordovensis* (De Leon). Only portions of holotype specimen were clear enough to show fine detail.

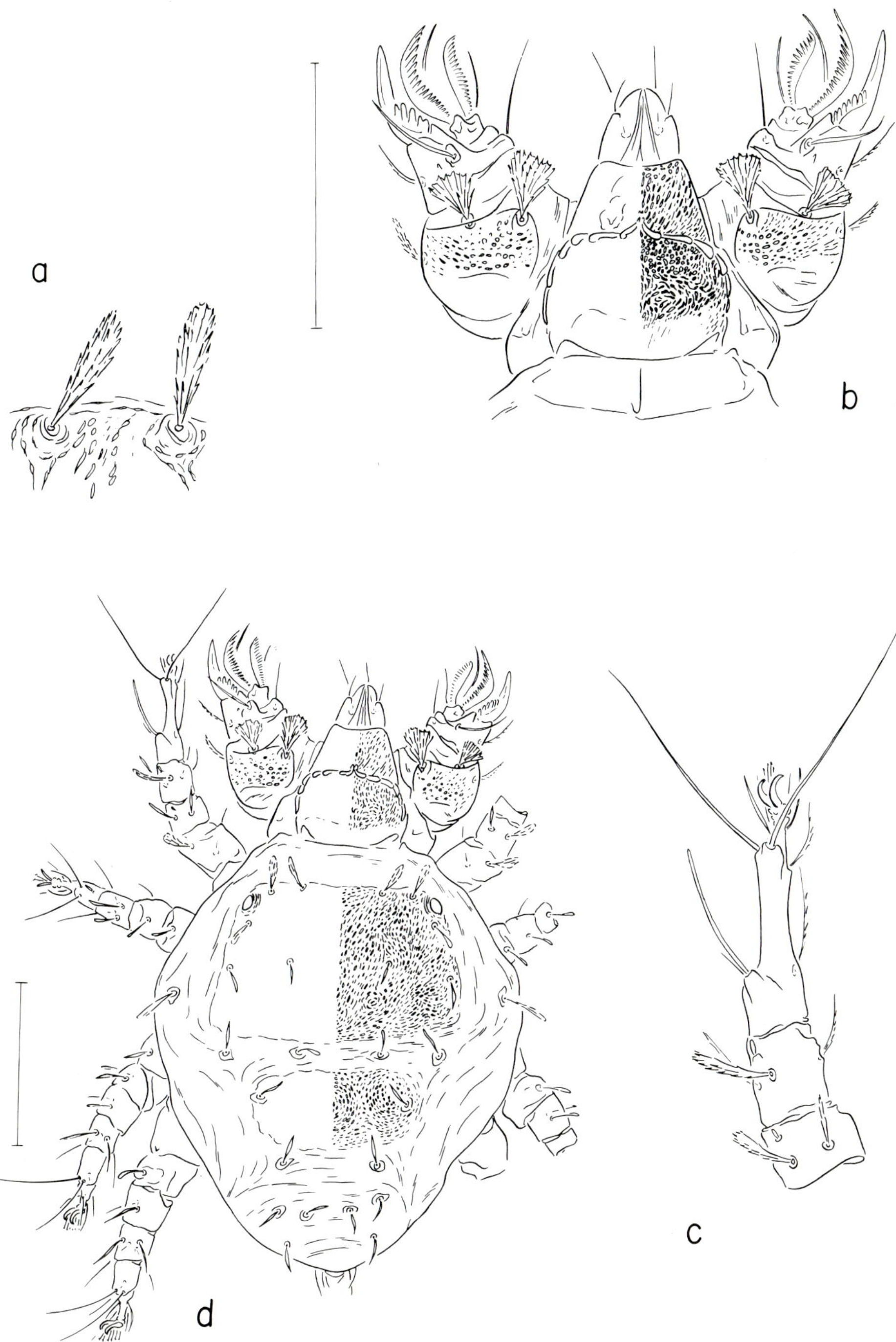

Fig. 9. *Hemicheyletia scutellata* (De Leon). *a*, First two dorsolateral propodosomal setae; *b*, gnathosoma; *c*, left leg 1; *d*, dorsal aspect of female.

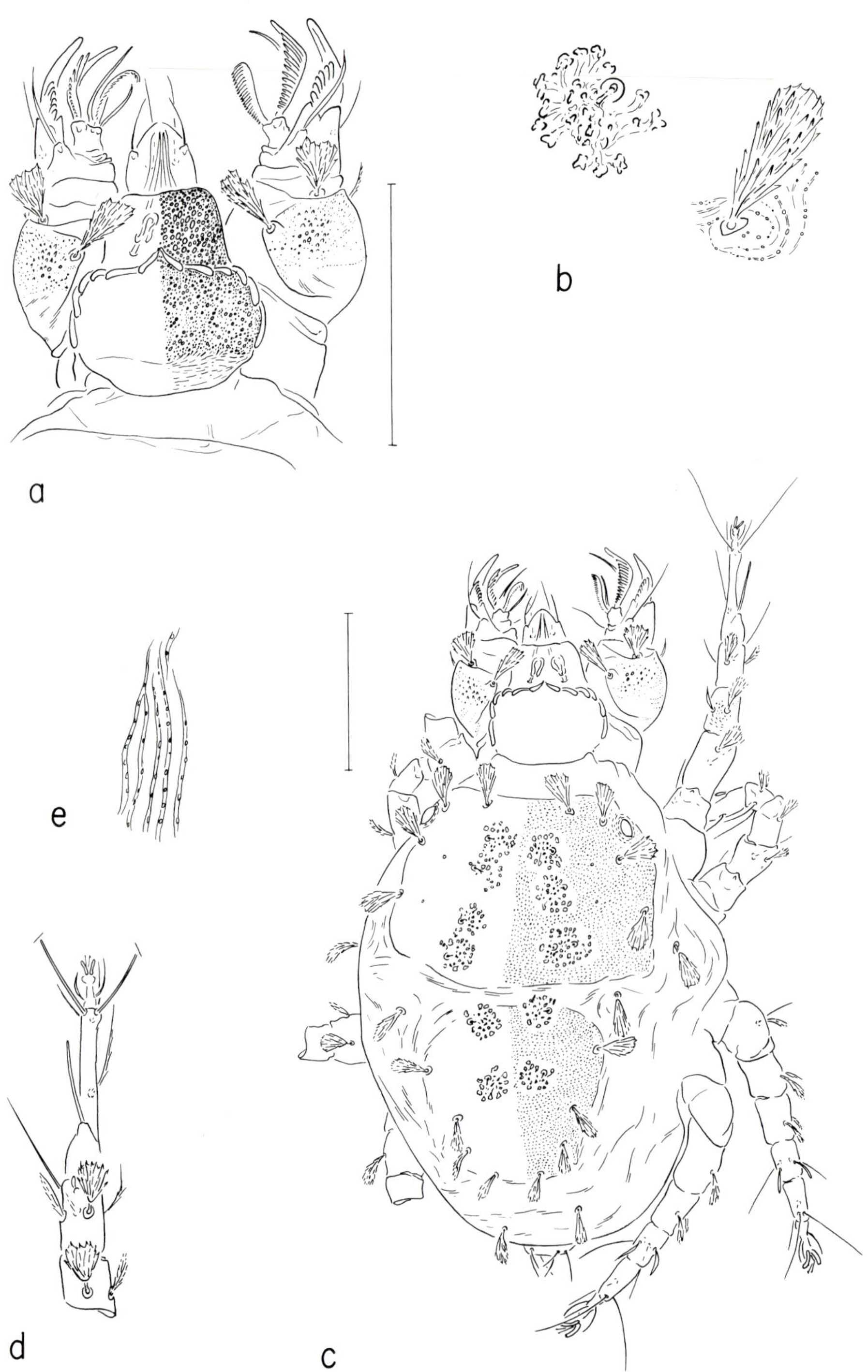

Fig. 10. *Hemicheyletia wellsi* (Baker). *a*, Gnathosoma; *b*, fragmented median seta (left) and first dorsolateral hysterosomal seta (right); *c*, dorsal aspect of female; *d*, left leg I; *e*, appearance of interscutal striae.

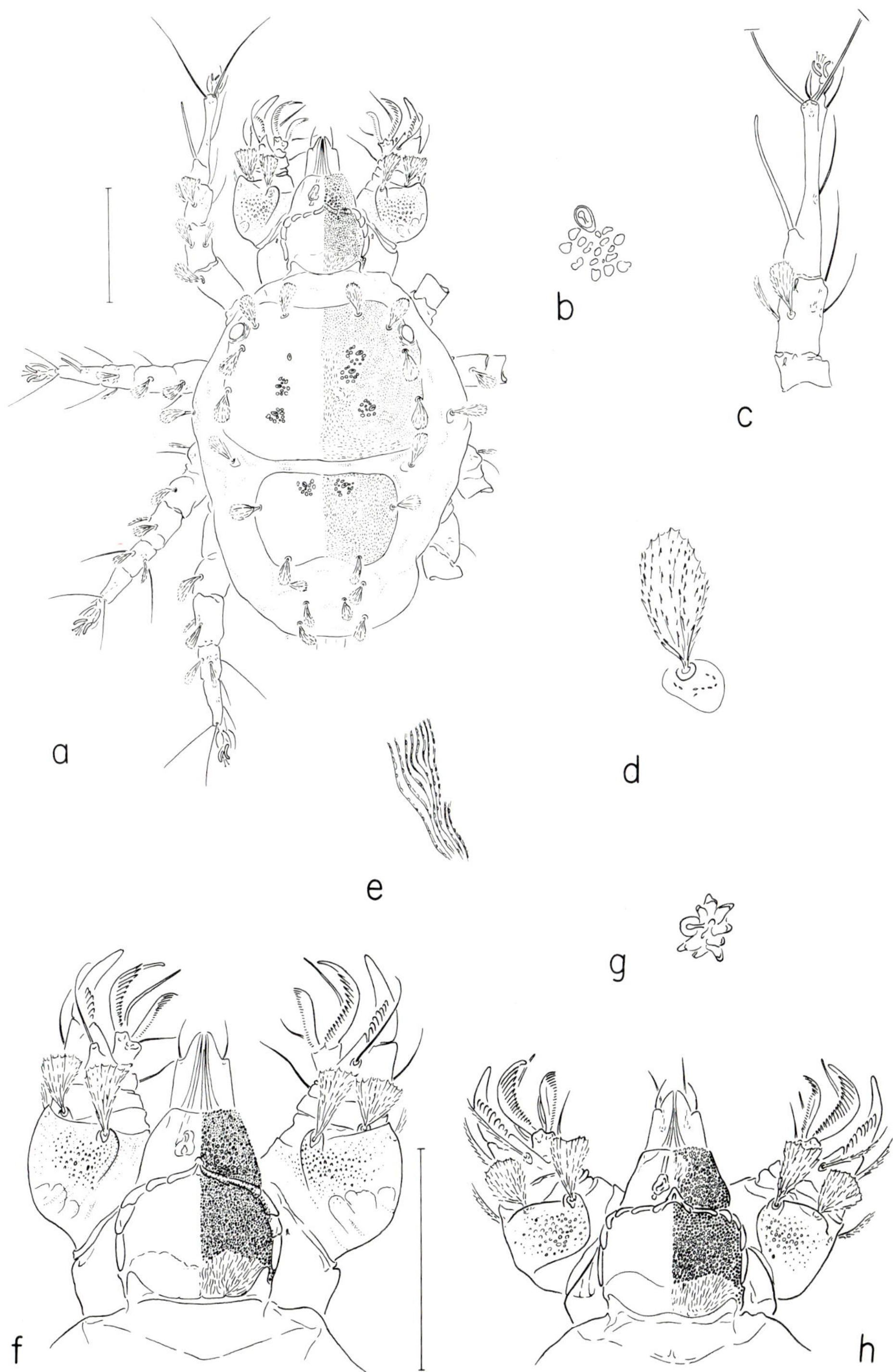

Fig. 11. *Hemicheyletia serrula*, n. sp. *a*, Dorsum, female; *b*, fragmented median seta; *c*, left leg I; *d*, first dorsolateral hysterosomal seta; *e*, several interscutal striae; *f*, gnathosoma. *Hemicheyletia darwinia*, n. sp. *g*, rosette-like dorsomedian seta; *h*, gnathosoma, female.

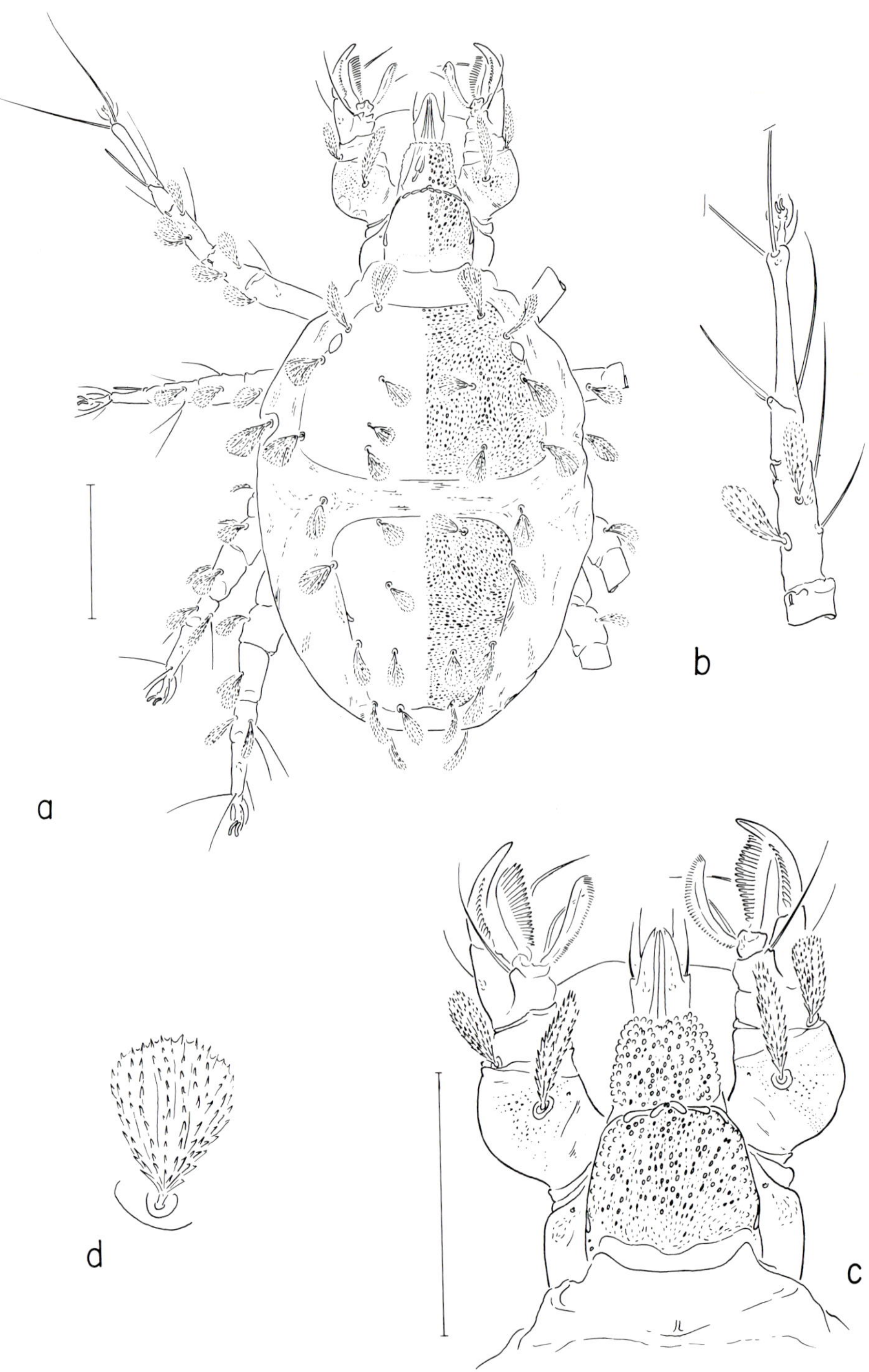

Fig. 12. *Hemicheyletia rostella*, n. sp. *a*, Dorsum, female; *b*, left leg I; *c*, gnathosoma; *d*, first dorsolateral hysterosomal seta.

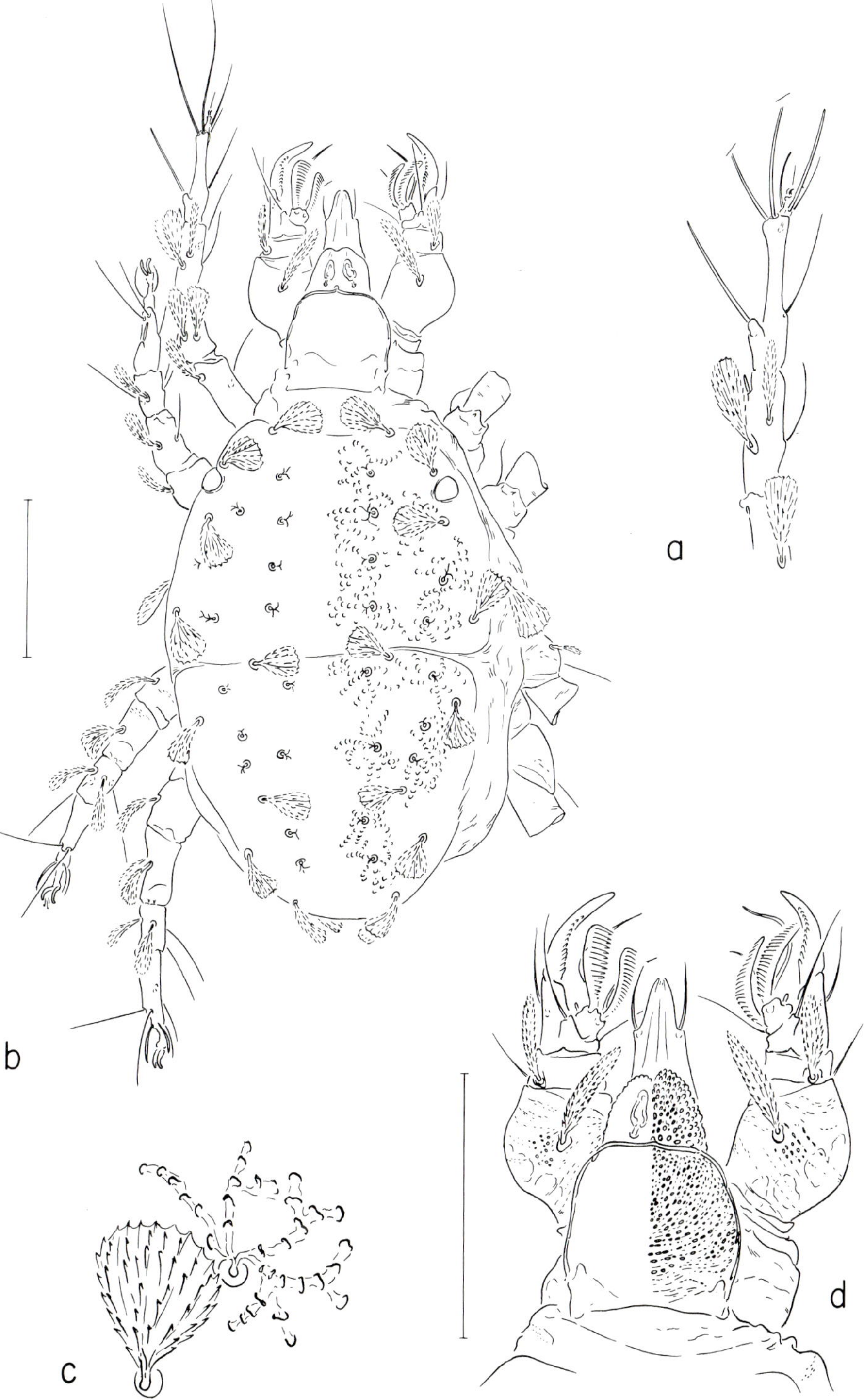

Fig. 13. *Paracheyletia pyriformis* (Banks). *a*, Left leg I; *b*, dorsum, female; *c*, second dorso-lateral hysterosomal seta and adjacent dorsomedian seta; *d*, gnathosoma.

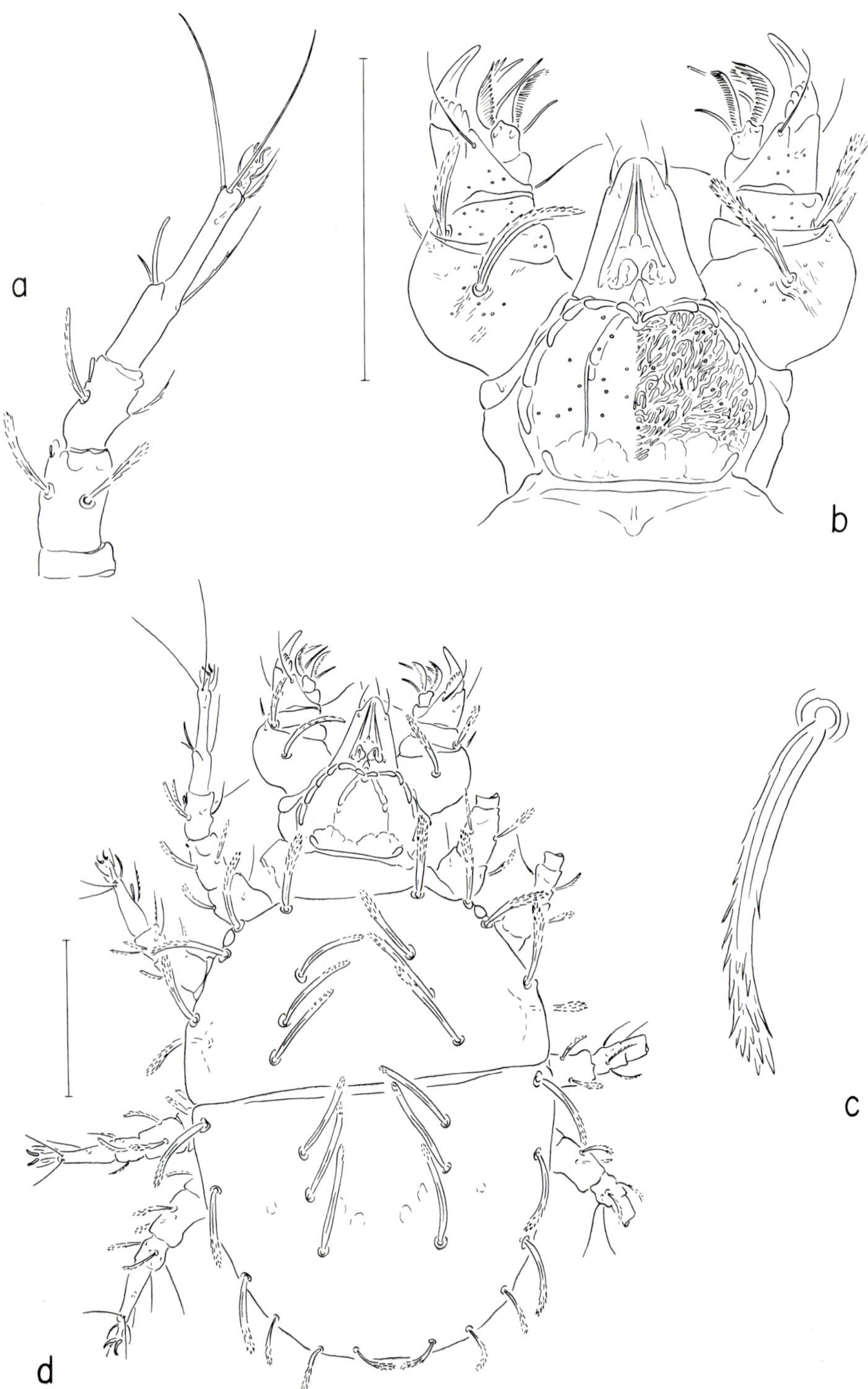

Fig. 14. *Laeliocheyletia teretis*, n. sp. *a*, Left leg I; *b*, gnathosoma; *c*, first dorsolateral hysterosomal seta; *d*, dorsal aspect of female.

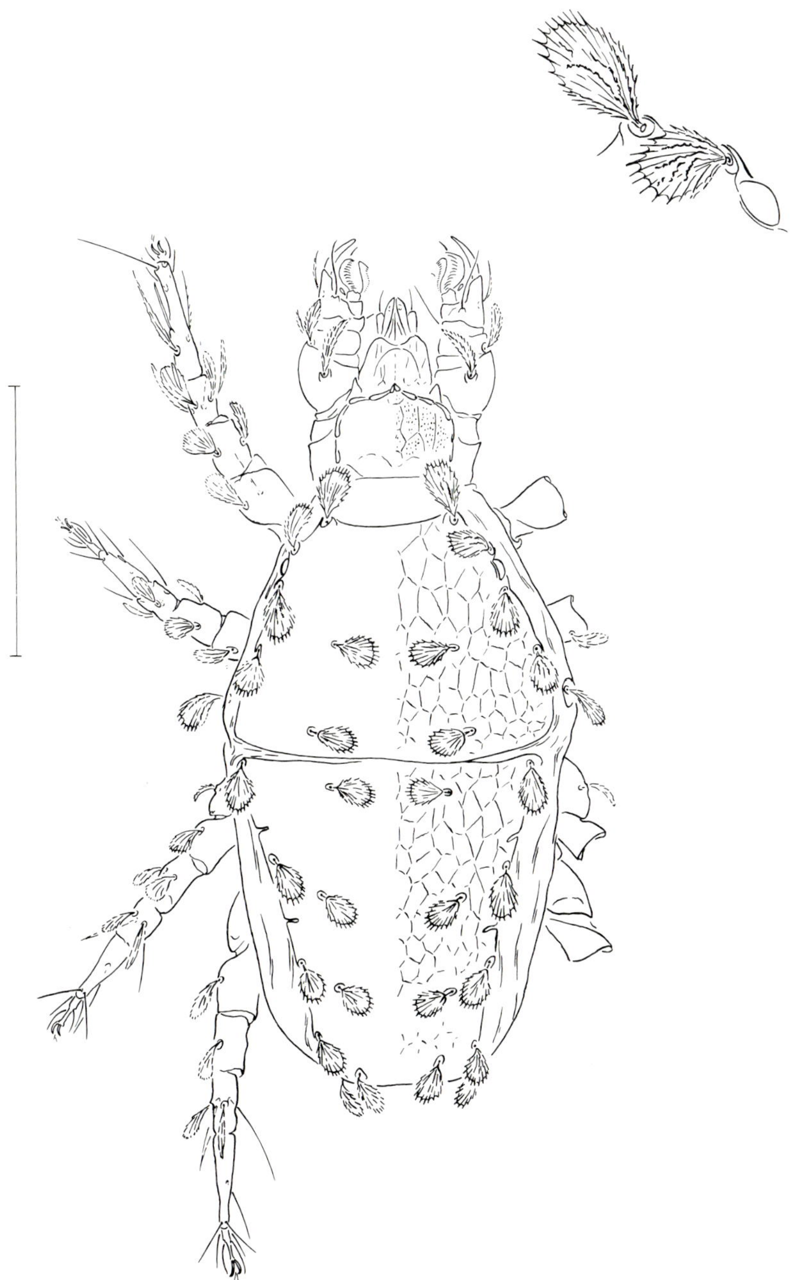

Fig. 15. *Ker palmatus* Muma. Dorsal aspect of female, with first two dorsolateral propodosomal setae drawn to larger scale.

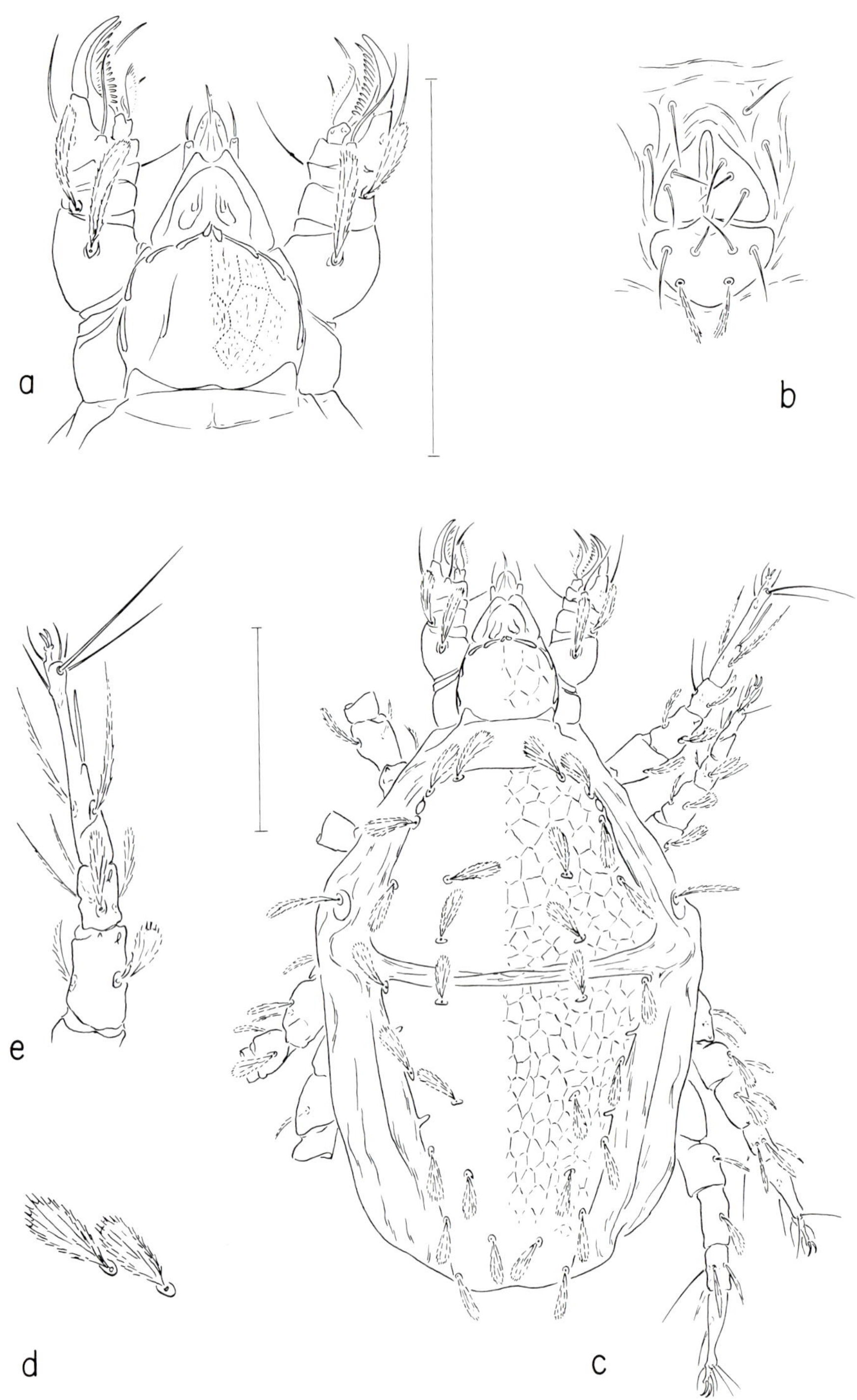

Fig. 16. *Ker bakeri* Zaher and Soliman. *a*, Gnathosoma; *b*, anogenital covers; *c*, dorsal aspect of female; *d*, first and second dorsolateral propodosomal setae; *e*, left leg I.

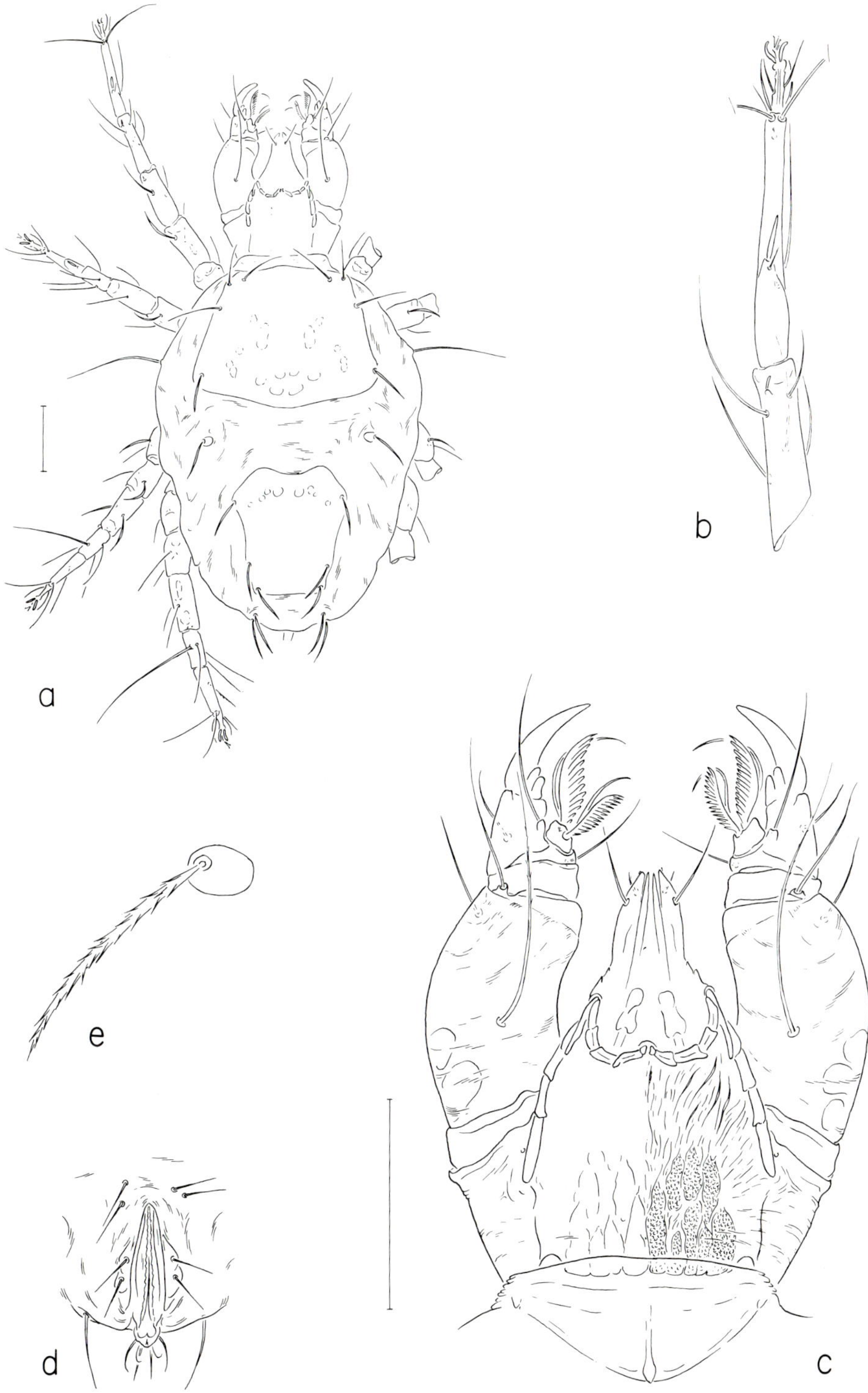

Fig. 17. *Cheyletus eruditus* (Schrank). *a*, Dorsal aspect of female; *b*, tarsus I with part of tibia; *c*, gnathosoma; *d*, anogenital region; *e*, first dorsolateral hysterosomal seta.

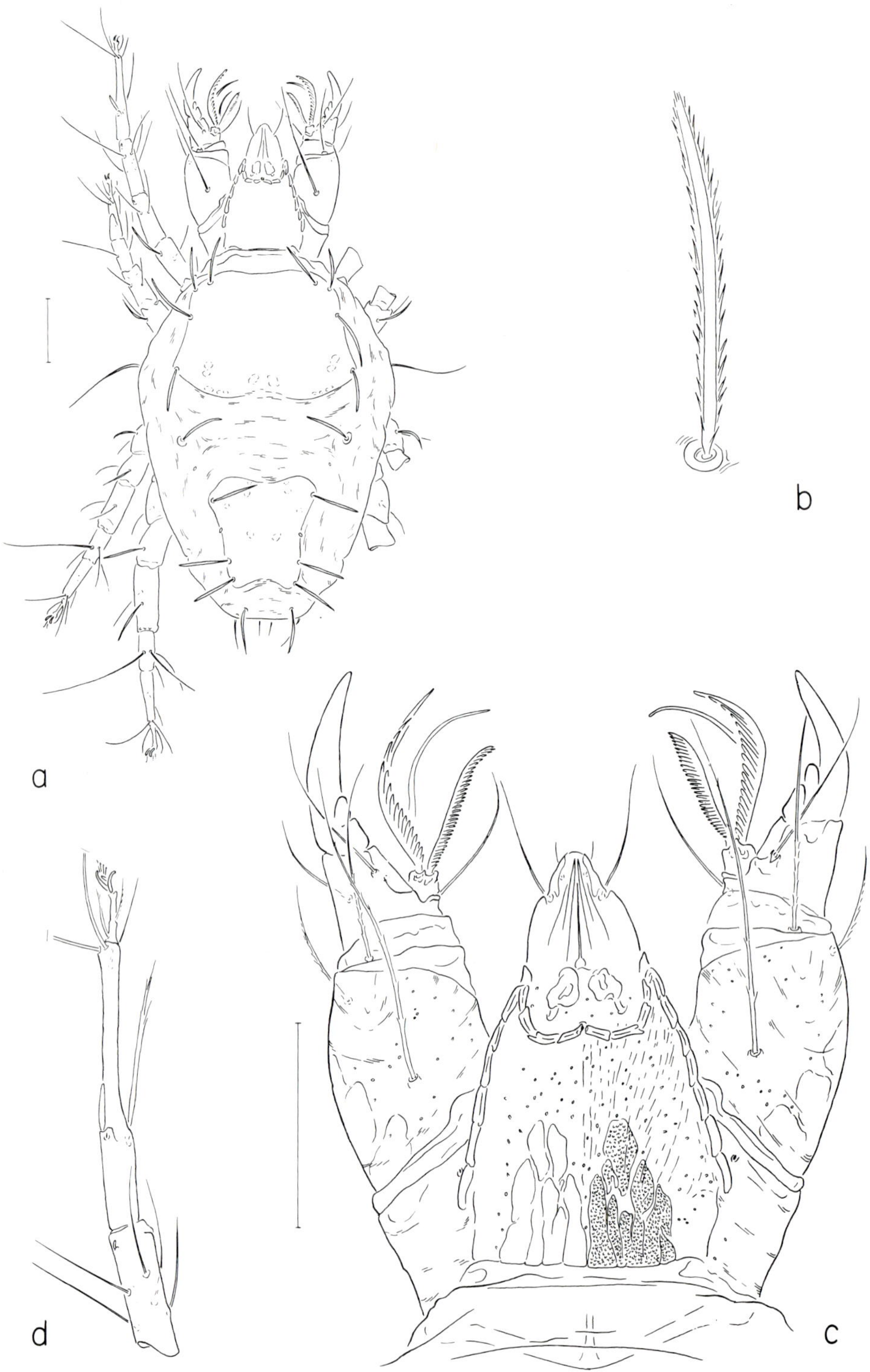

Fig. 18. *Cheyletus malaccensis* Oudemans. *a,* Female, dorsum; *b,* first dorsolateral hysterosomal seta; *c,* gnathosoma; *d,* left tarsus I with part of tibia.

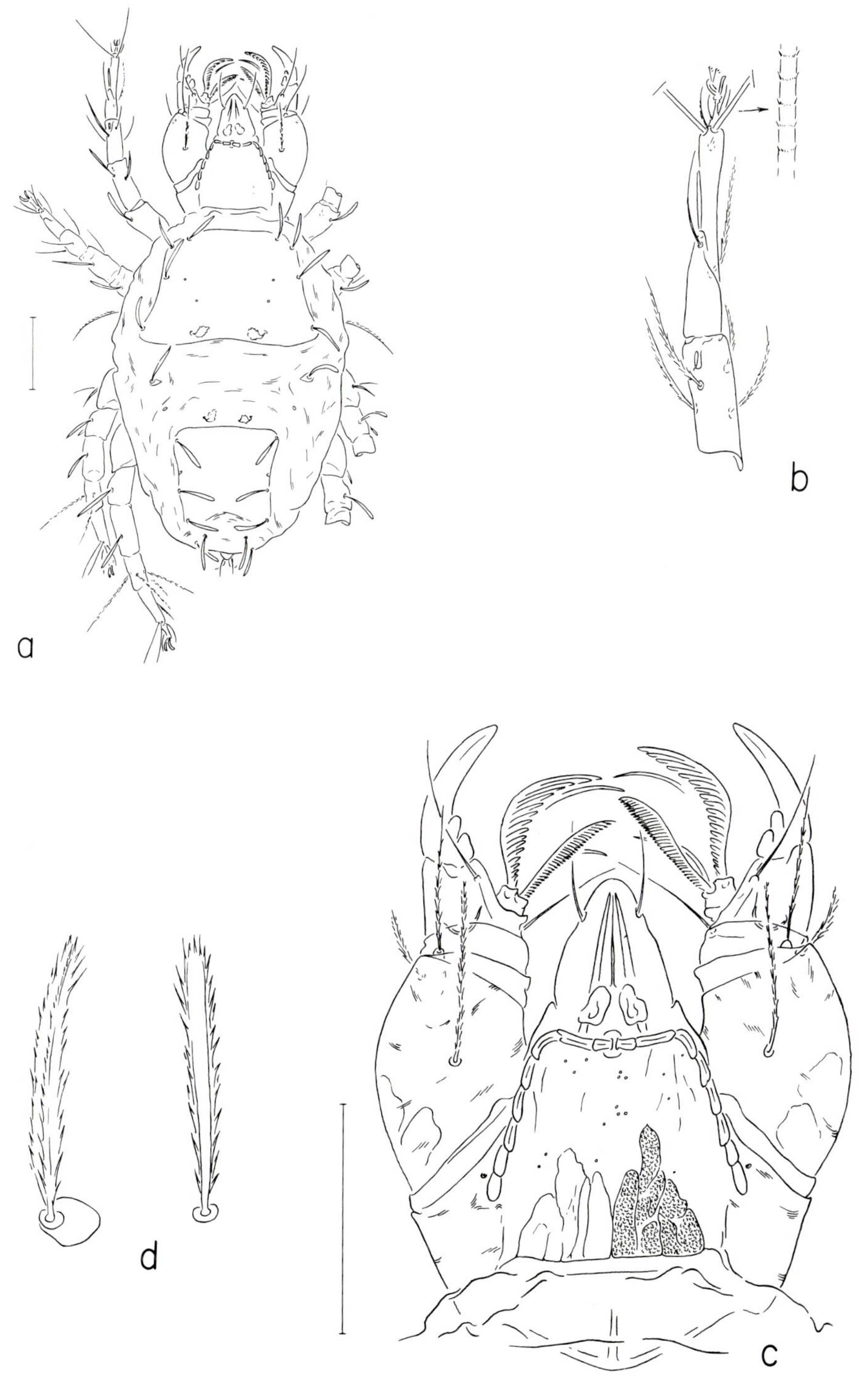

Fig. 19. *Cheyletus aversor* Rohdendorf. *a*, Female, dorsum; *b*, tarsus and part of tibia, left leg 1; *c*, gnathosoma; *d*, first (left) and second (right) setae of hysterosoma.

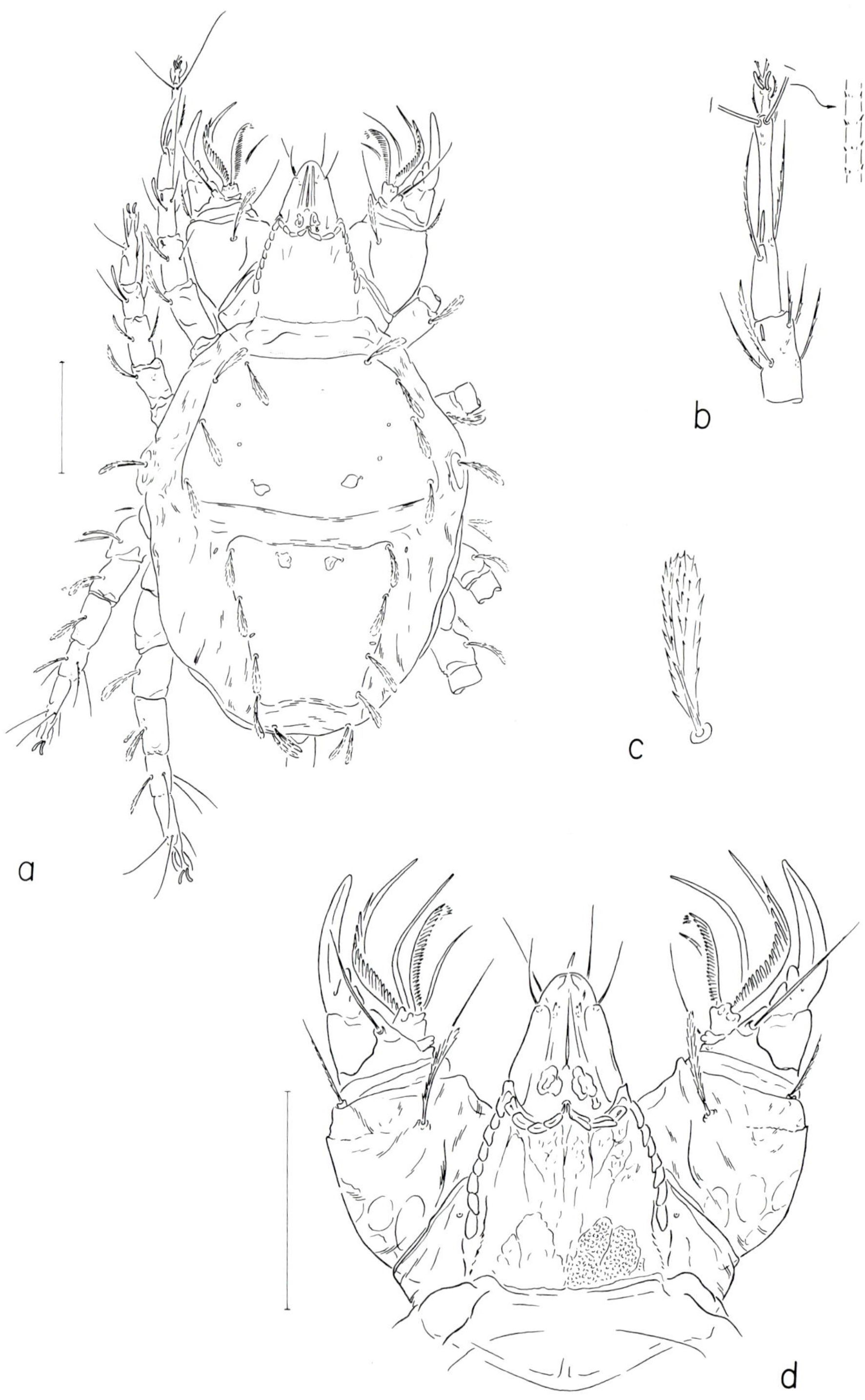

Fig. 20. *Cheyletus cacahuamilpensis* Baker. *a*, Female, dorsum; *b*, part of left leg ɪ; *c*, first dorsolateral hysterosomal seta; *d*, gnathosoma.

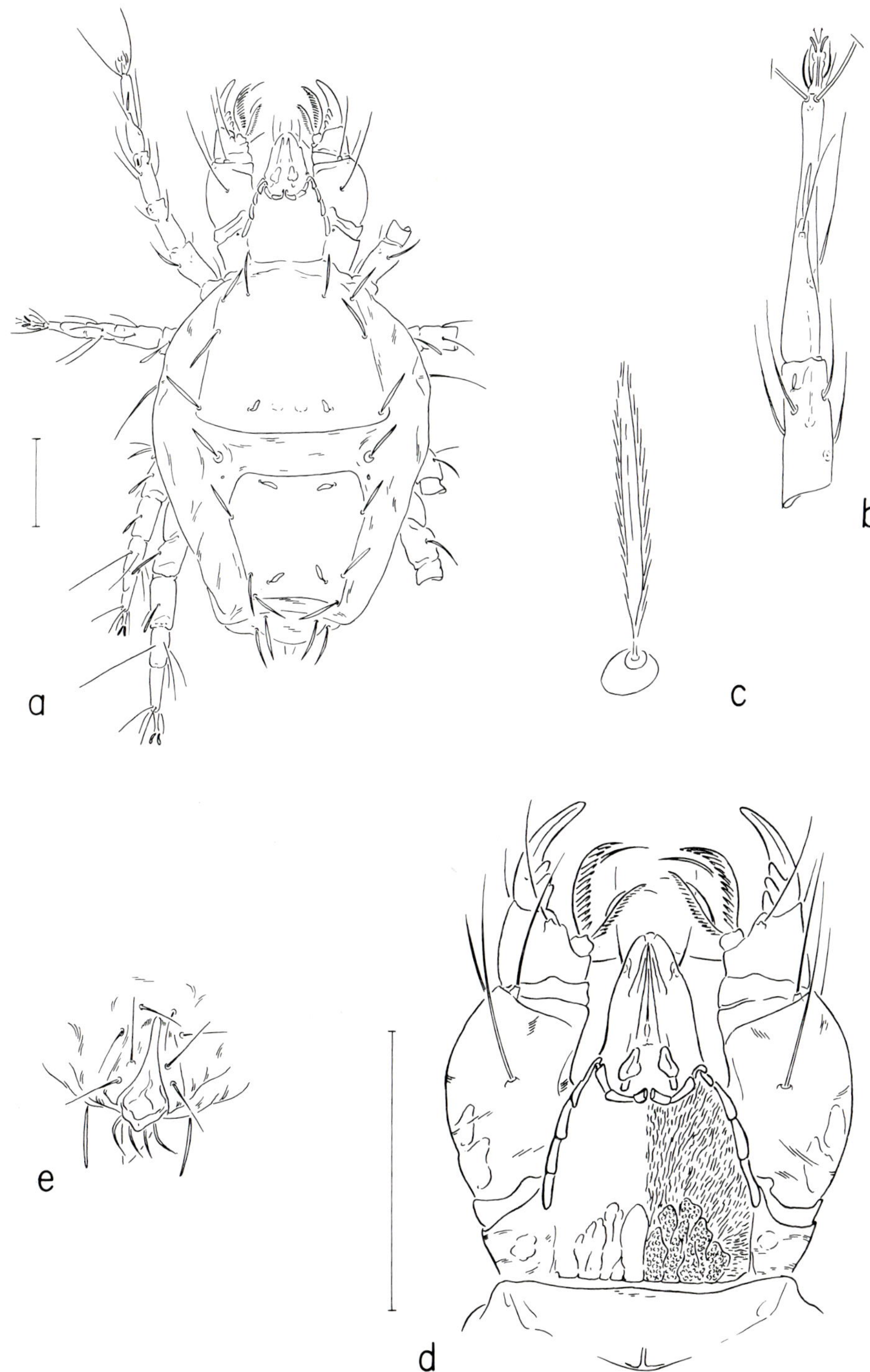

Fig. 21. *Cheyletus trouessarti* Oudemans. *a*, Dorsal aspect of female; *b*, part of left leg I; *c*, first dorsolateral hyterosomal seta; *d*, gnathosoma; *e*, anogenital region.

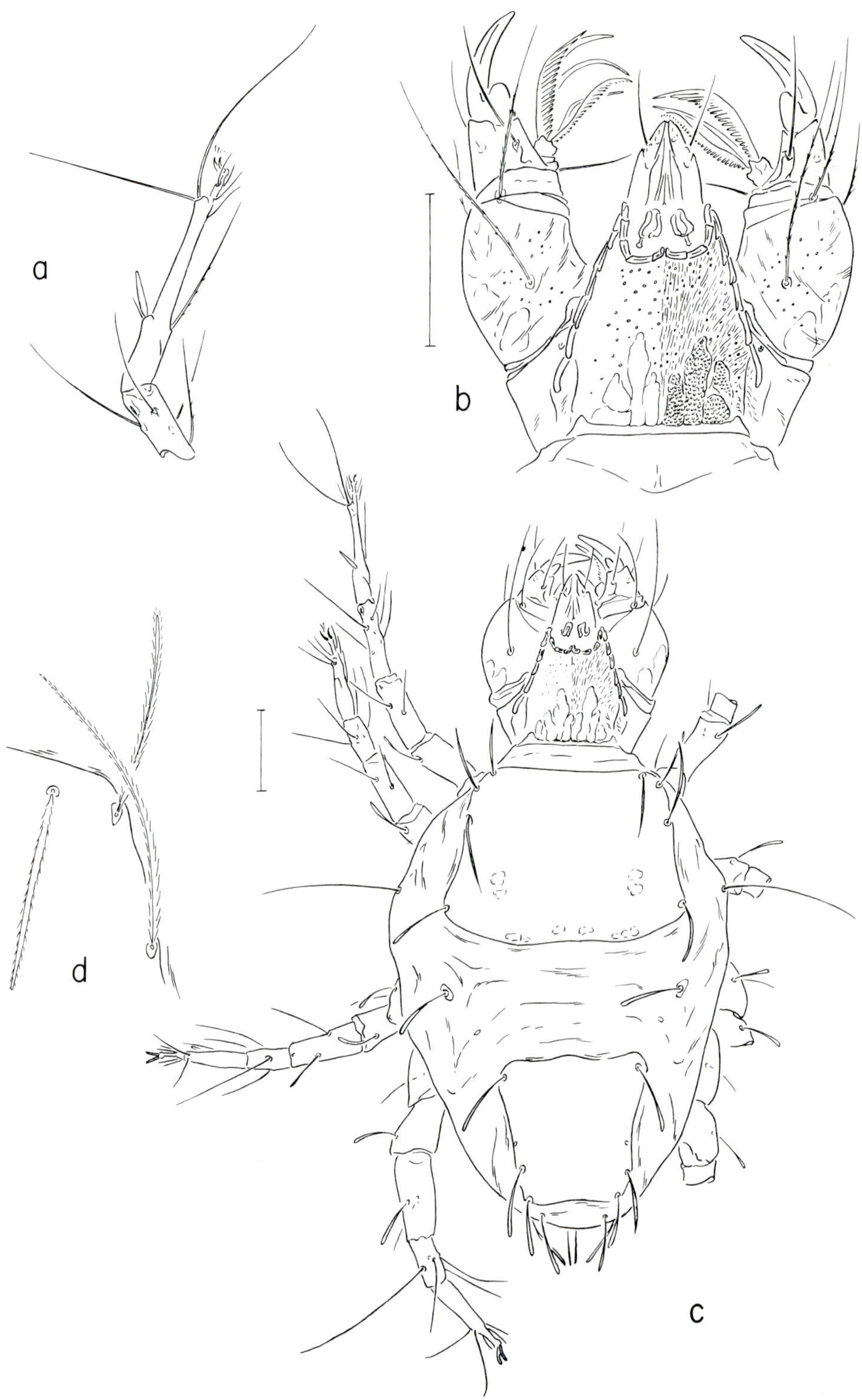

Fig. 22. *Cheyletus fortis* Oudemans. *a*, Tarsus and tibia, left leg I; *b*, gnathosoma; *c*, female, dorsum; *d*, first three dorsolateral setae of propodosoma.

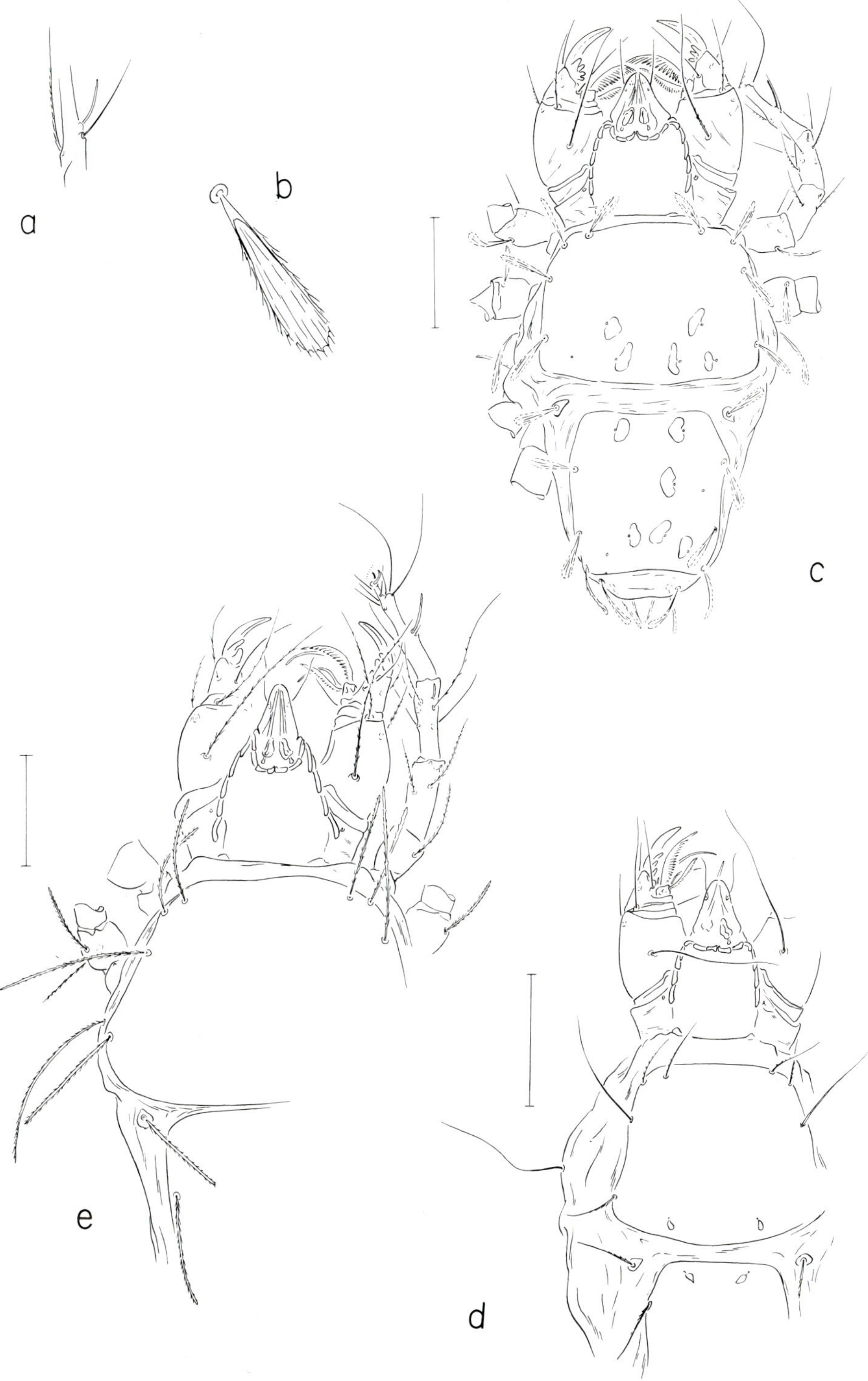

Fig. 23. *Cheyletus linsdalei* Baker. *b*, Second dorsolateral hysterosomal seta; *c*, dorsal aspect of female. *Cheyletus hendersoni* Baker. *d*, female dorsum; *a*, midsection of tarsus I to show relations of *w*I and guard seta. *Cheyletus malayensis* Cunliffe. *e*, dorsal aspect of female.

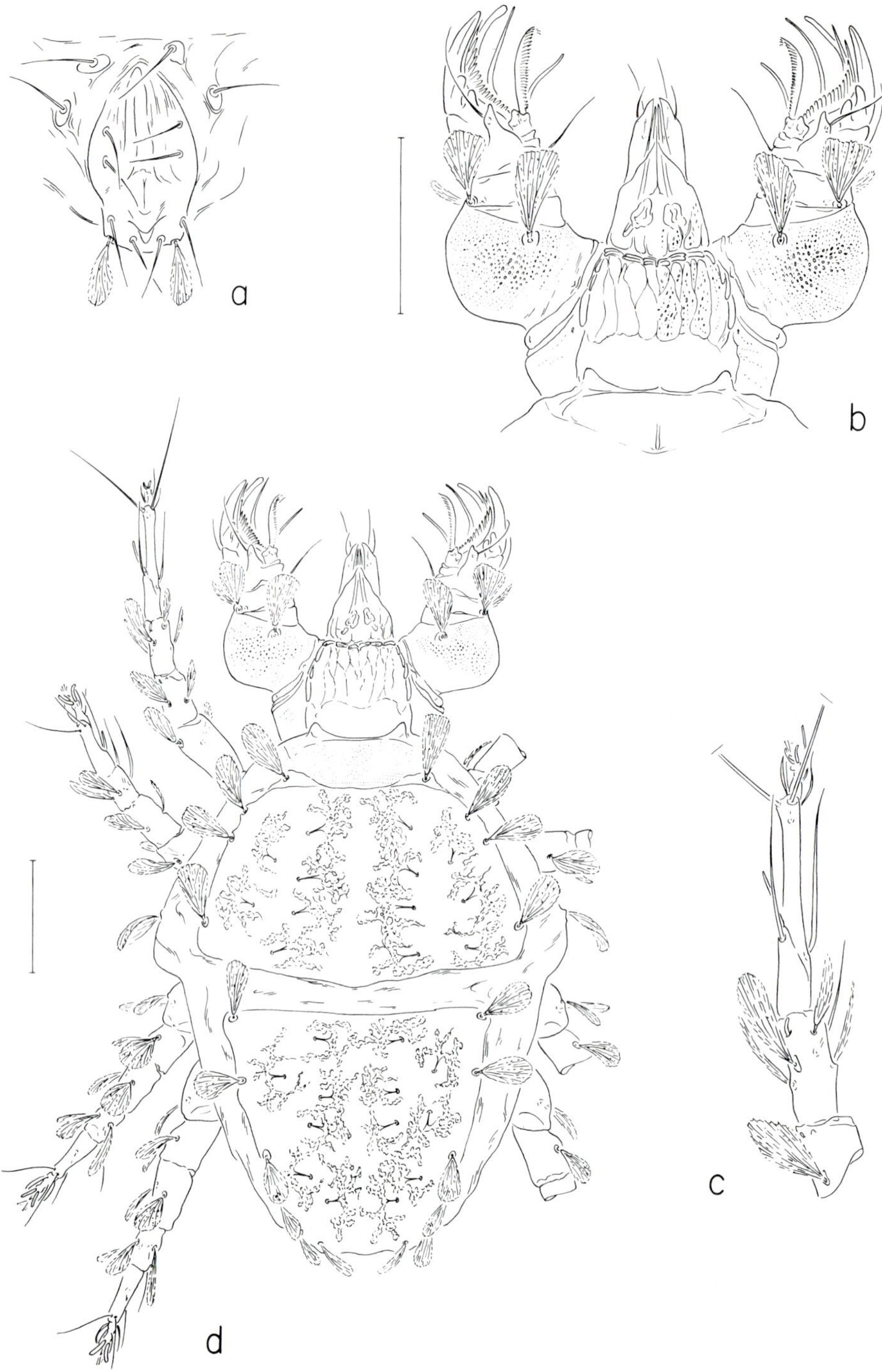

Fig. 24. *Eucheyletia bishoppi* Baker. *a*, Anogenital region; *b*, gnathosoma; *c*, part of left leg I; *d*, dorsum, female.

Fig. 25. *Eucheyletia sinensis* Volgin. *a*, Fourth dorsolateral propodosomal seta with dorsomedian seta nearby; *b*, gnathosoma; *c*, part of right leg I; *d*, female, dorsal aspect.

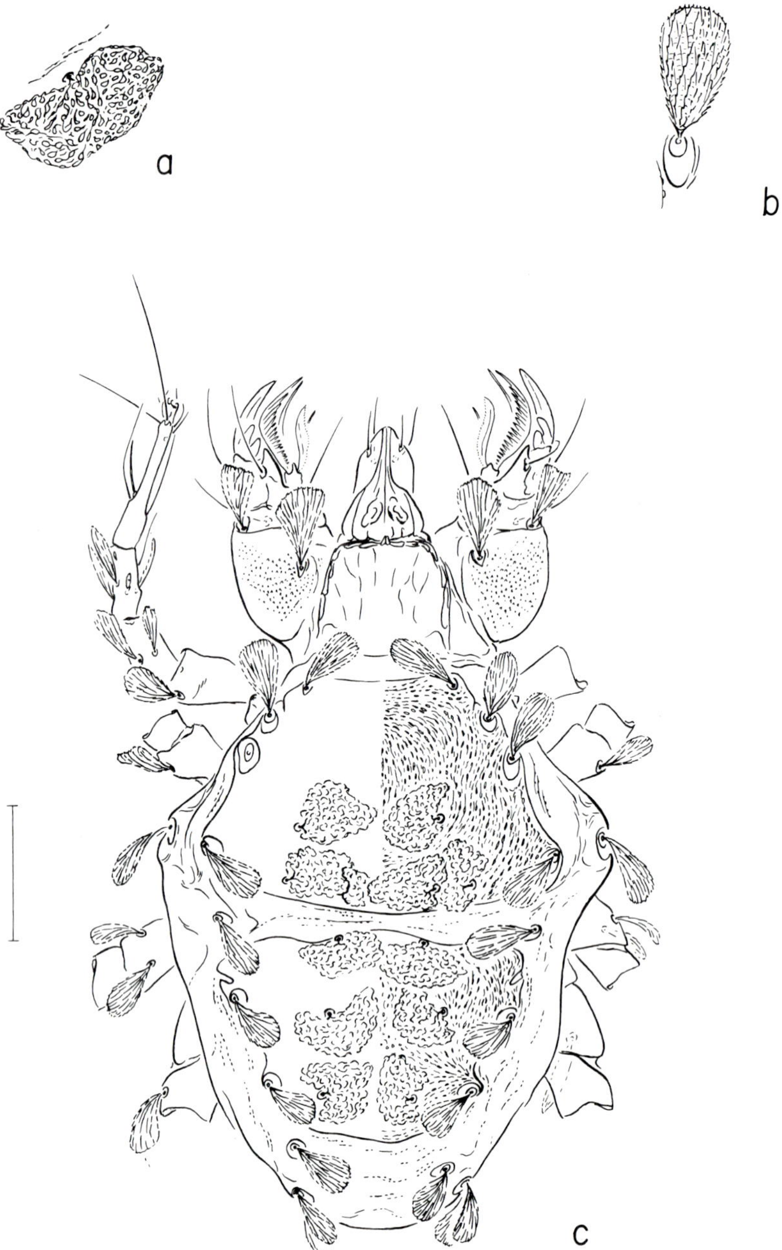

Fig. 26. *Eucheyletia hardyi* Baker. *a*, First dorsomedian seta on hysterosoma; *b*, third dorsolateral propodosomal seta; *c*, female, dorsum.

Fig. 27. *Eucheyletia reticulata* Cunliffe. *a*, Gnathosoma; *b*, first and second dorsolateral propodosomal setae; *c*, dorsal aspect of female; *d*, distal part of right leg I.

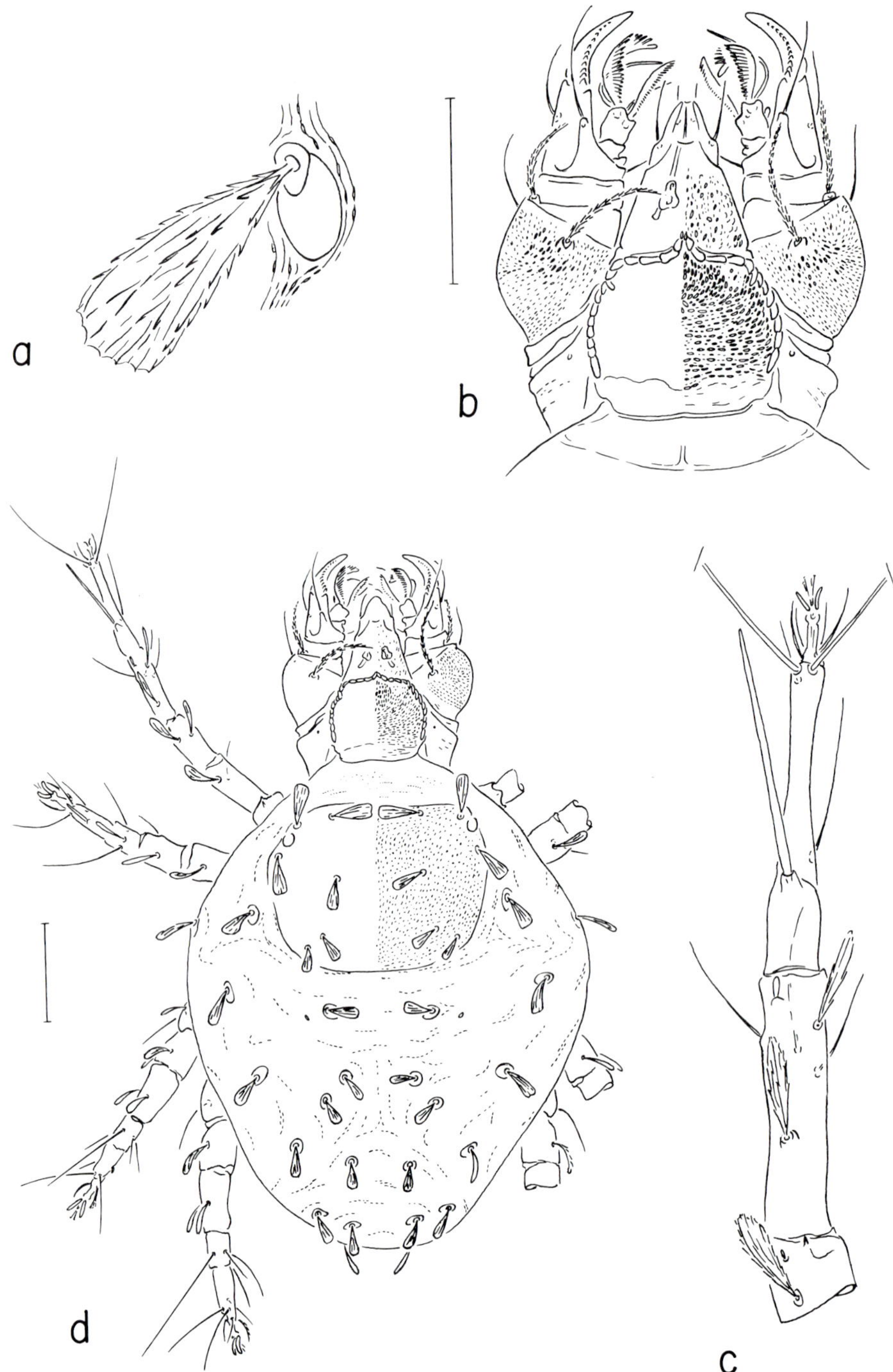

Fig. 28. *Cheletacarus gryphus*, n. sp. *a*, Anterolateral hysterosomal seta. *b*, gnathosoma; *c*, tibia and tarsus of left leg I; *d*, dorsal aspect of female.

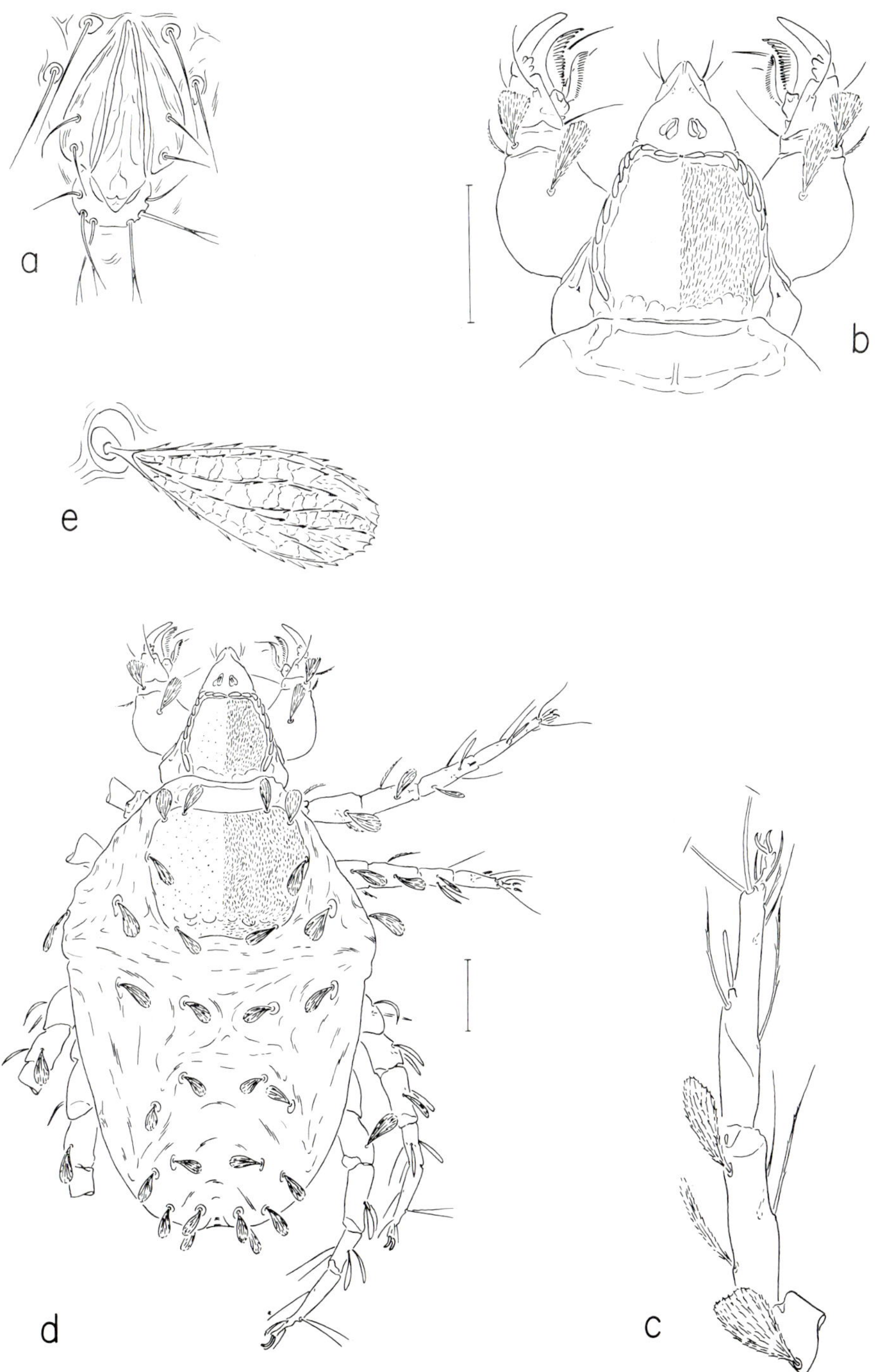

Fig. 29. *Cheletonella vespertilionis* Womersley. *a*, Anogenital region; *b*, gnathosoma; *c*, left leg I; *d*, dorsum, female; *e*, first dorsolateral hysterosomal seta.

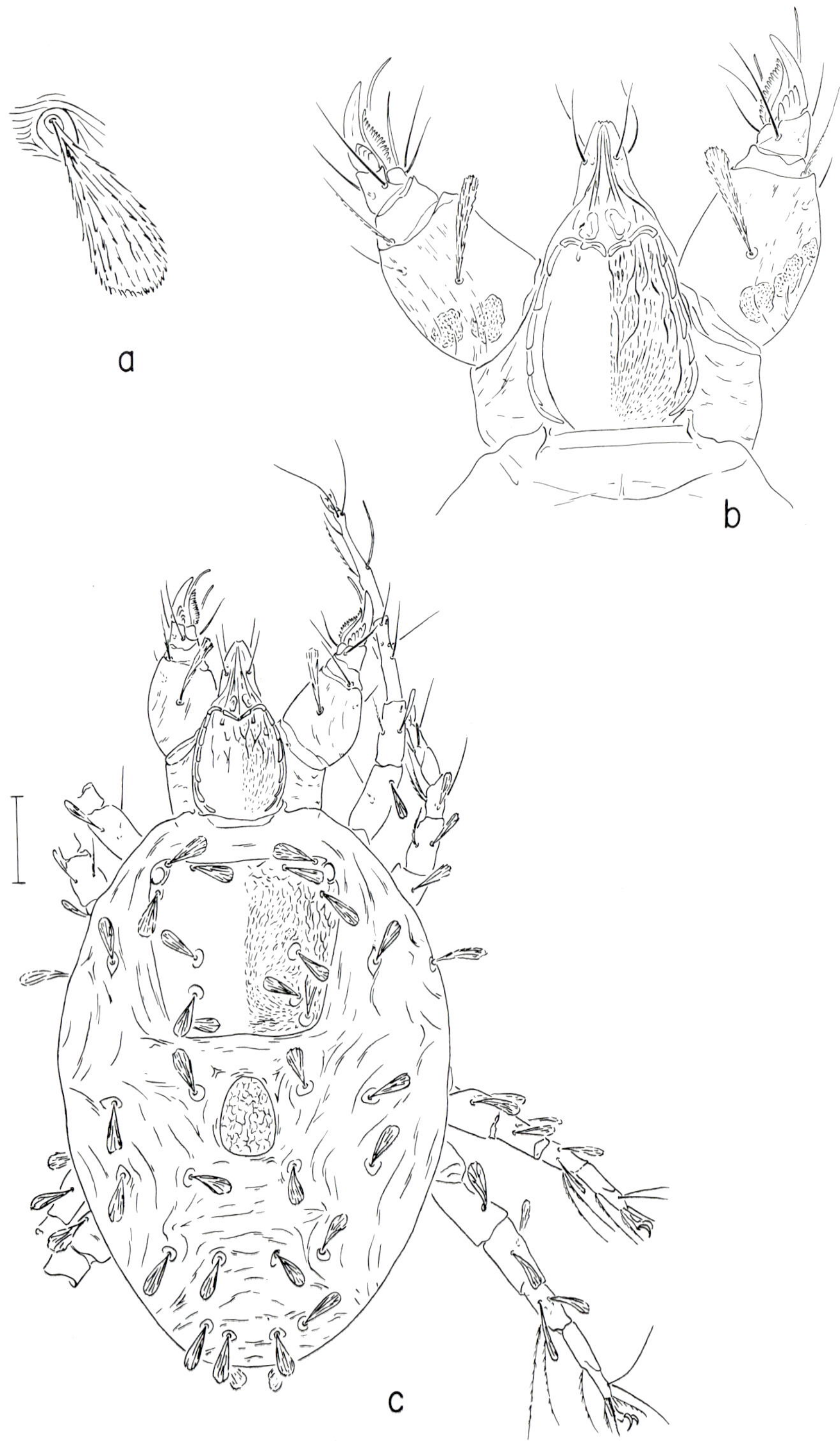

Fig. 30. *Cheletonata milesi* Womersley. *a*, First dorsolateral hysterosomal seta; *b*, gnathosoma; *c*, upper side, female.

Fig. 31. *Paracaropsis travisi* (Baker). Gnathosoma and dorsal aspect of holotype specimen.

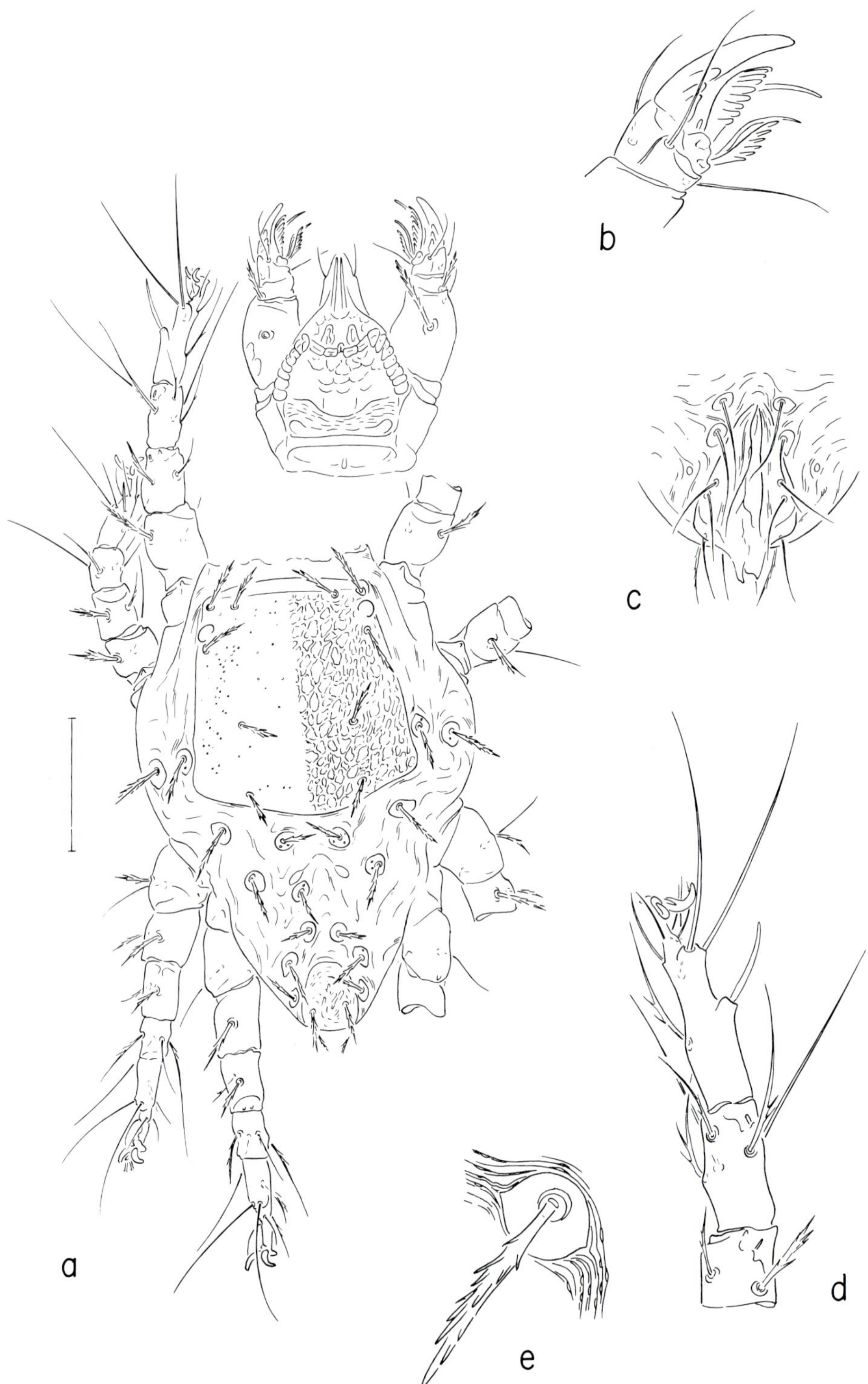

Fig. 32. *Cheletophyes eckerti,* n. sp. *a,* Idiosoma with gnathosoma separated, dorsal; *b,* tip of pedipalp; *c,* anogenital region; *d,* right leg I; *e,* first dorsolateral hysterosomal seta.

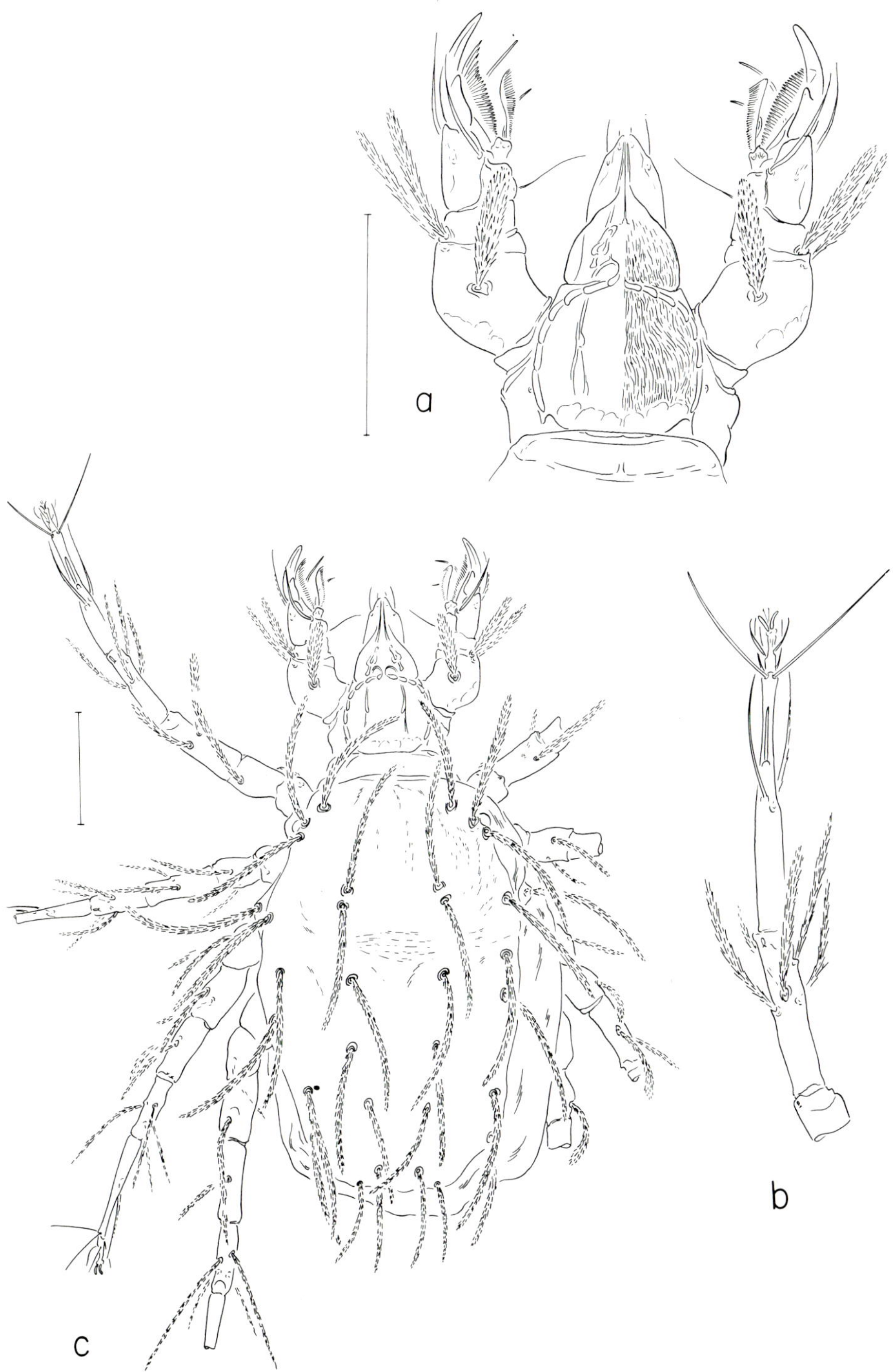

Fig. 33. *Nodele calamondin* Muma. *a*, Gnathosoma; *b*, left leg 1; *c*, dorsum, female.

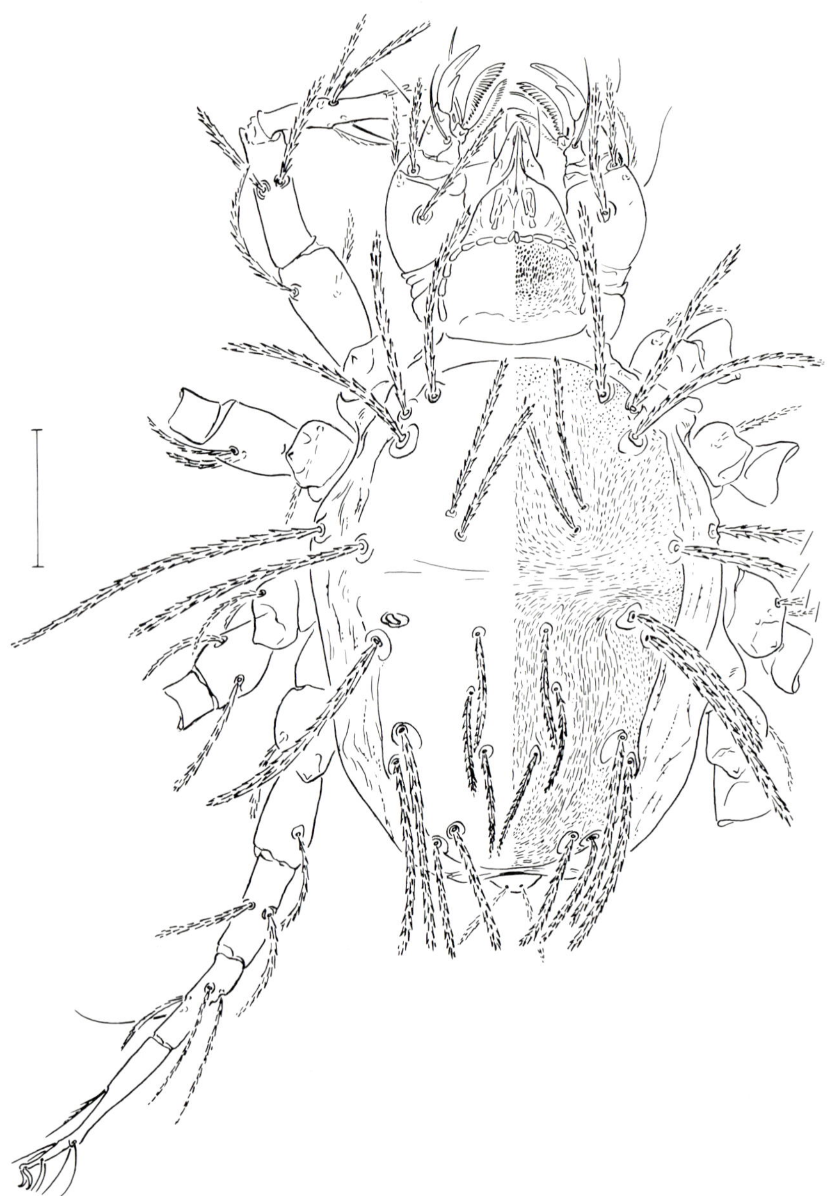

Fig. 34. *Nodele philippinensis* (Baker).

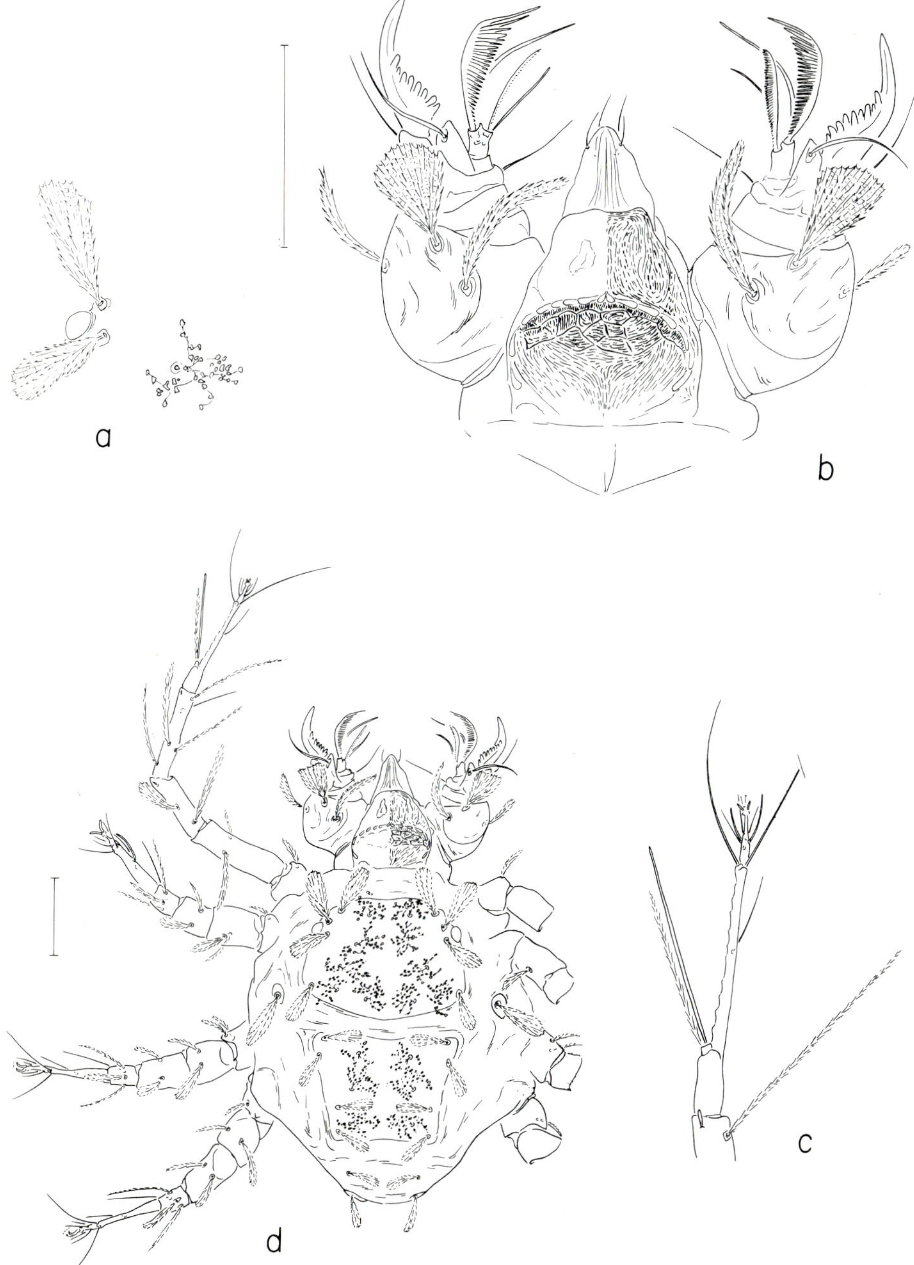

Fig. 35. *Mexecheles cunliffei* De Leon. *a*, Second and third dorsolateral propodosomal setae with one dorsomedian seta nearby; *b*, gnathosoma; *c*, left tarsus I; *d*, dorsal side of female.

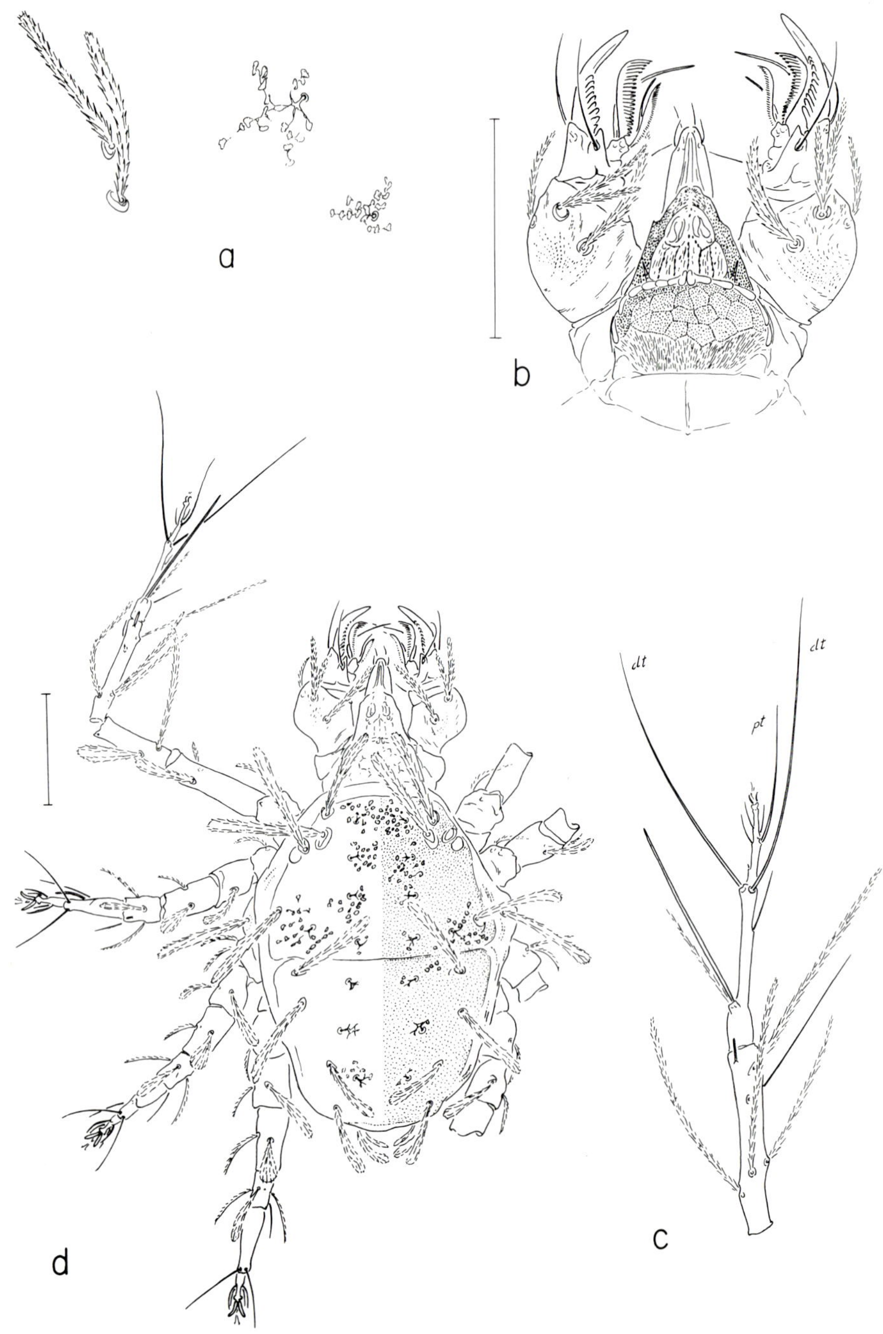

Fig. 36. *Mexecheles hawaiiensis* (Baker). *a*, First and second dorsolateral and two dorsomedian setae of propodosoma; *b*, gnathosoma; *c*, left leg I; *d*, female, dorsal aspect.

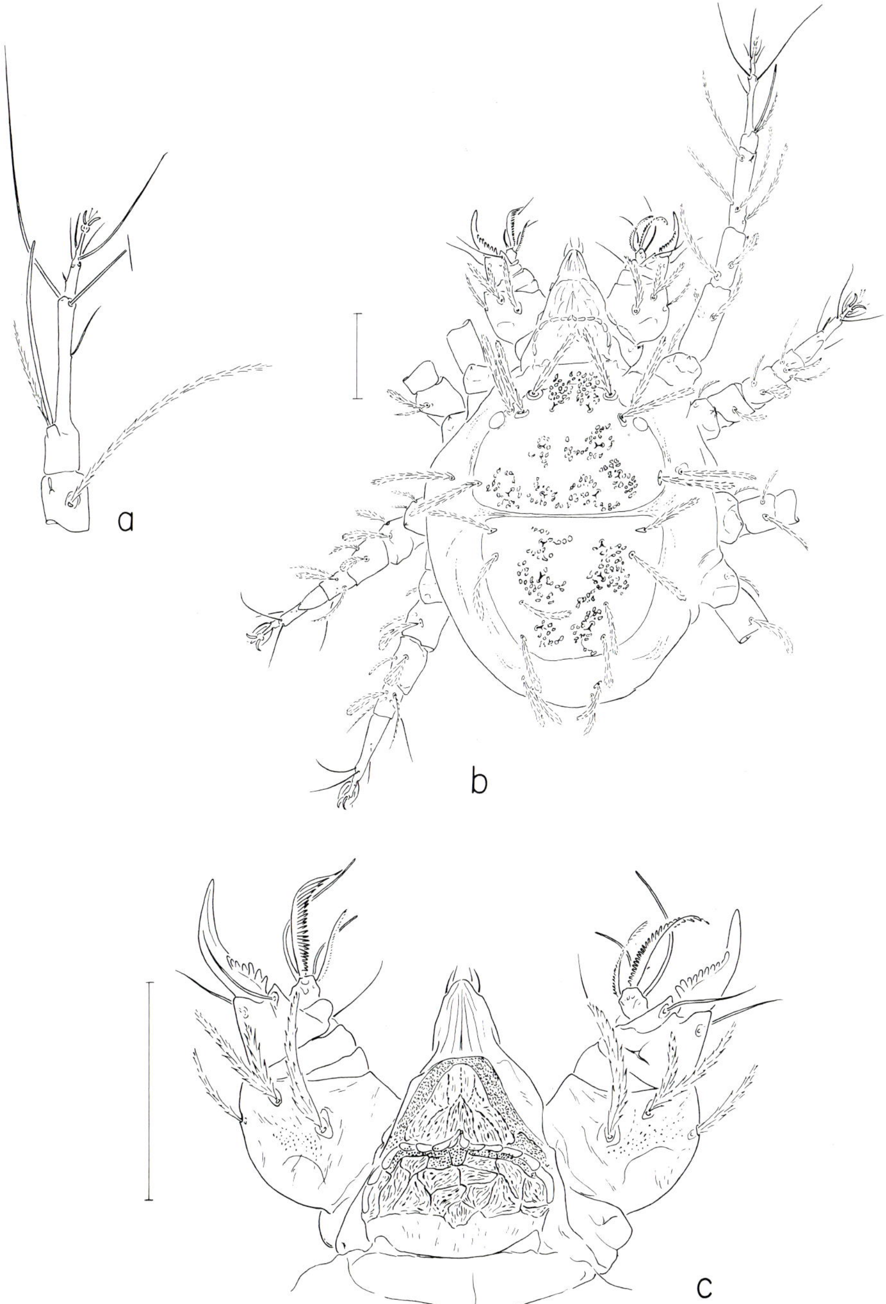

Fig. 37. *Mexecheles aztecorum* De Leon. *a*, Part of left leg I; *b*, dorsal aspect of female; *c*, gnathosoma.

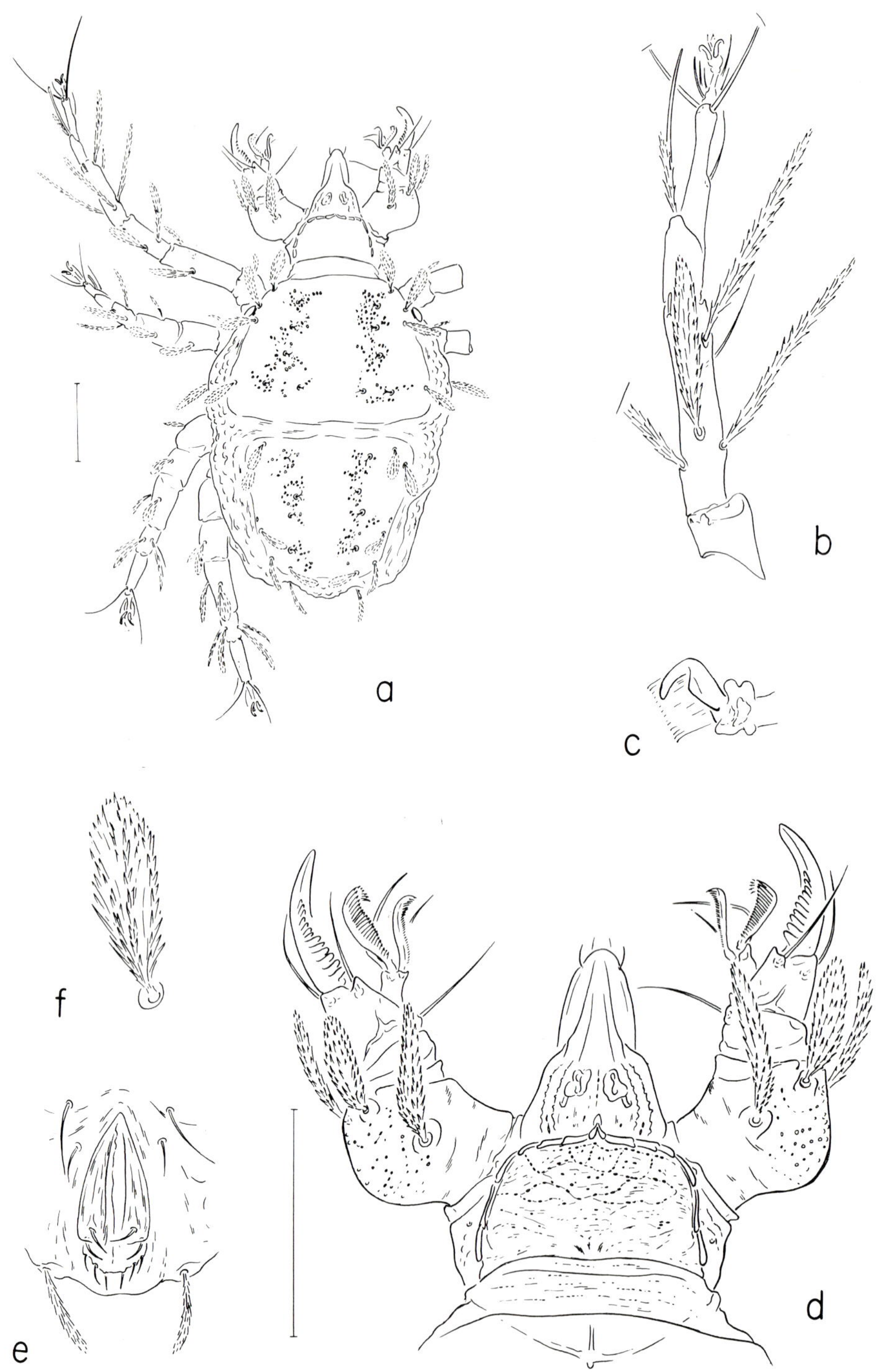

Fig. 38. *Mexecheles panneus*, n. sp. *a*, Female, dorsum; *b*, left tarsus and tibia I; *c*, processes on claws of leg IV; *d*, gnathosoma; *e*, anogenital region; *f*, first dorsolateral hysterosomal seta.

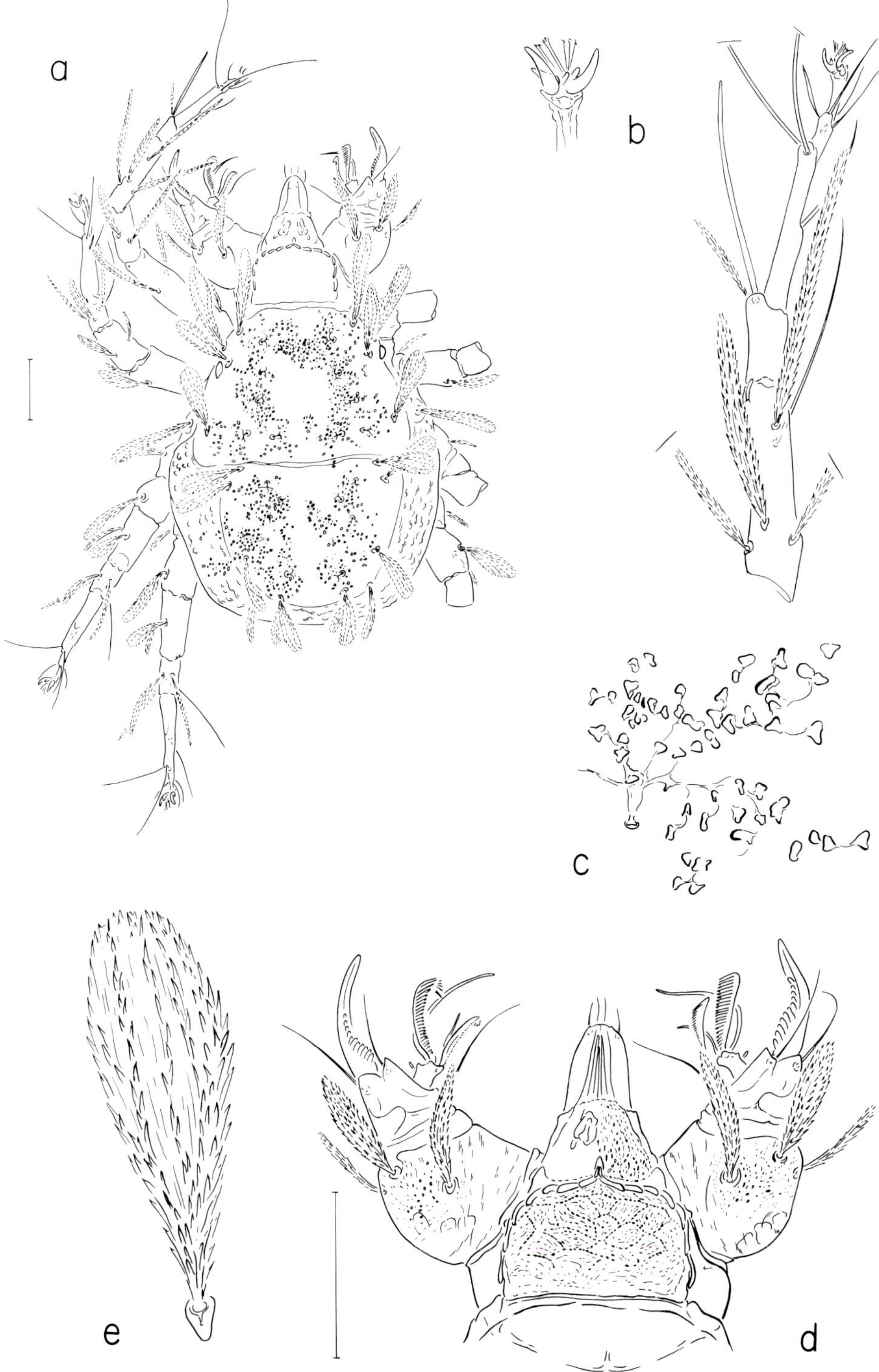

Fig. 39. *Mexecheles virginiensis* (Baker). *a*, Entire mite from above; *b*, tarsus and tibia I, with enlarged view of pretarsus I to illustrate ungual apophyses; *c*, one median seta of hysterosoma; *d*, gnathosoma; *e*, anteriormost dorsolateral seta of hysterosoma.

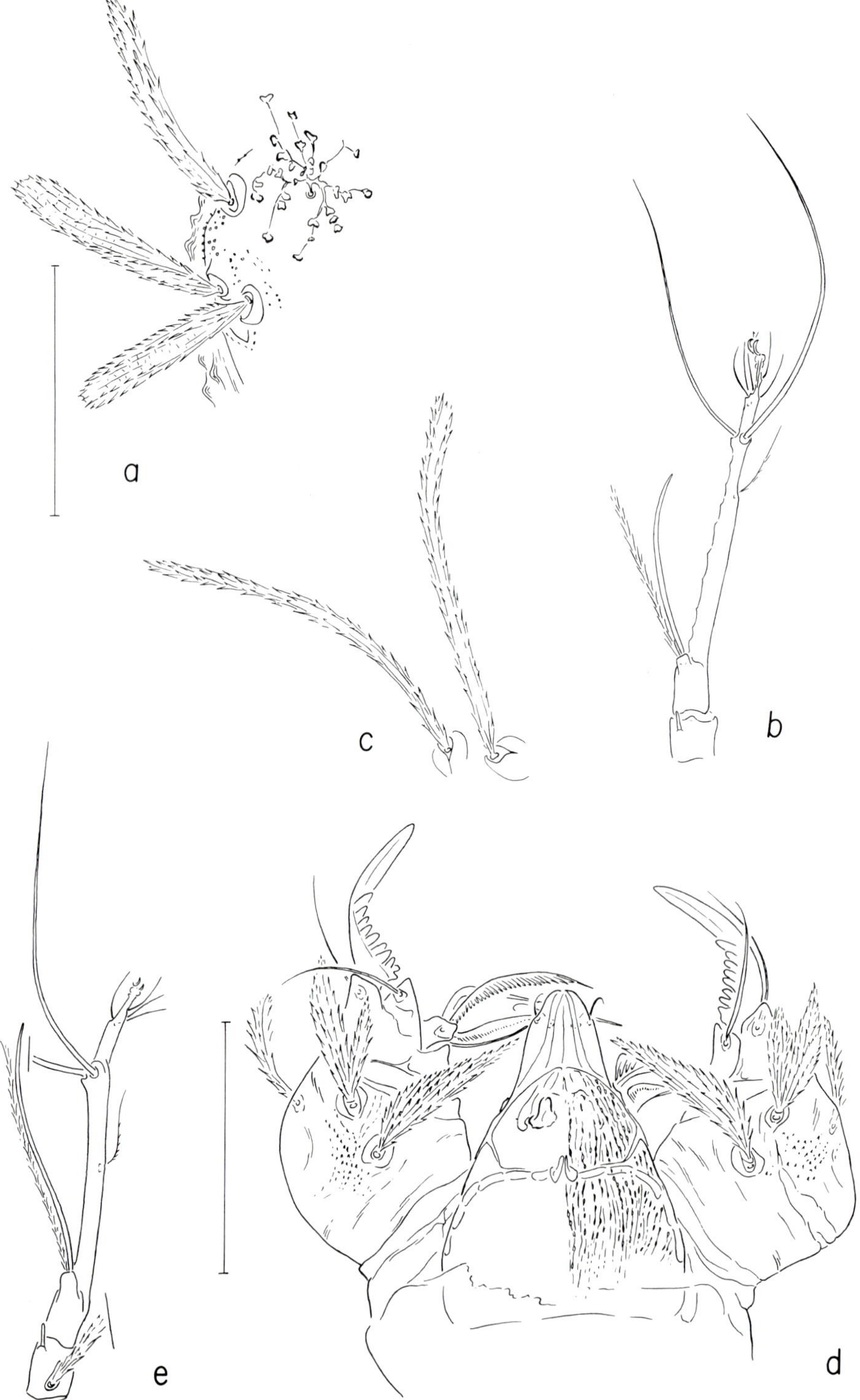

Fig. 40. *Mexecheles impolitus* (Smiley and Moser). *a*, Portion of propodosoma with first three dorsolateral setae and first median seta; *b*, left tarsus I. *Mexecheles marshalli* (Baker). *c*, First and second dorsolateral hysterosomal setae; *d*, gnathosoma; *e*, left tarsus I.

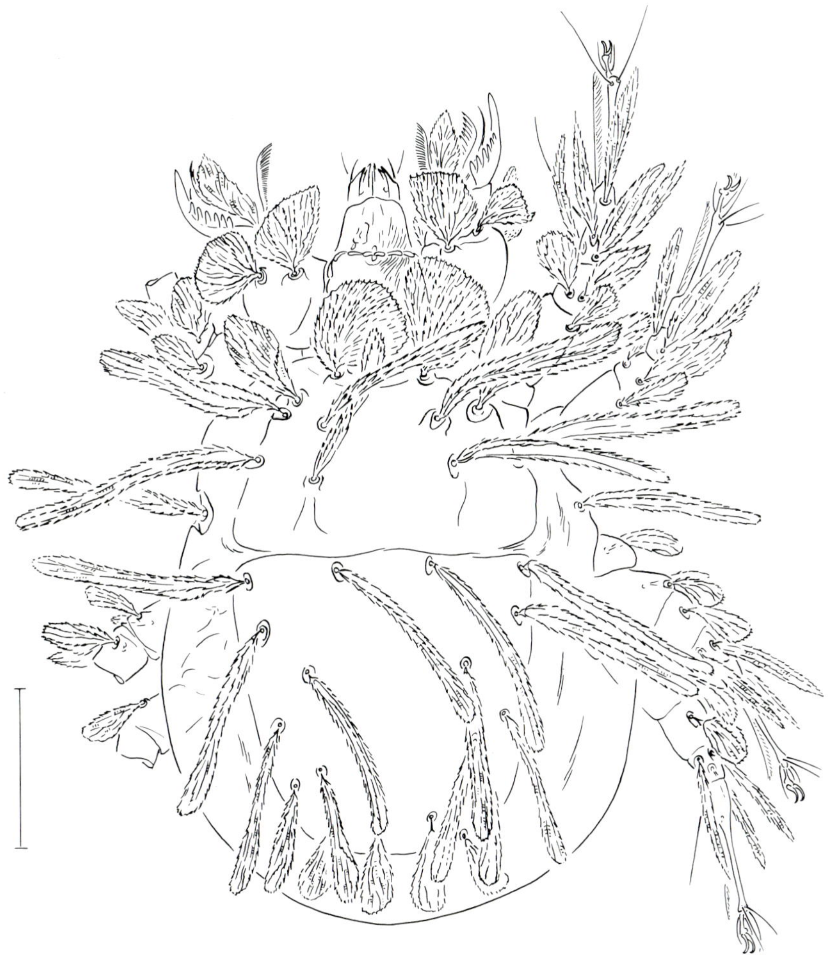

Fig. 41. *Grallacheles bakeri* De Leon. Dorsal aspect of female.

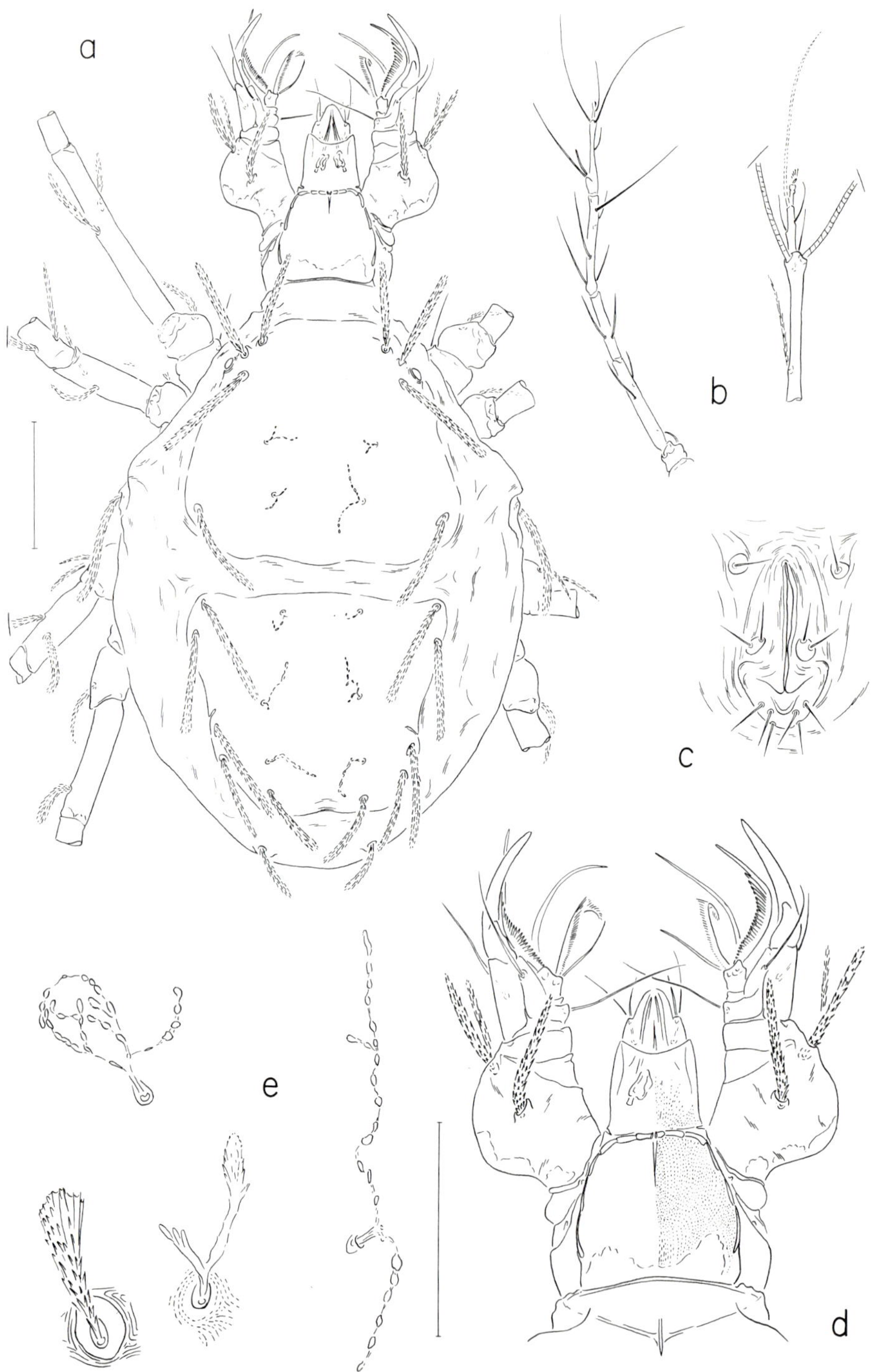

Fig. 42. *Cheletomorpha lepidopterorum* (Shaw). *a*, Female, dorsum; *b*, leg I and distal end of tarsus I; *c*, anogenital region; *d*, gnathosoma; *e*, median setae variously developed: orthodox condition in deutonymph (lower left); rare condition of incomplete metamorphic change in seta observed on adult female (lower middle); fully modified median setae commonly present on females (right and upper figures).

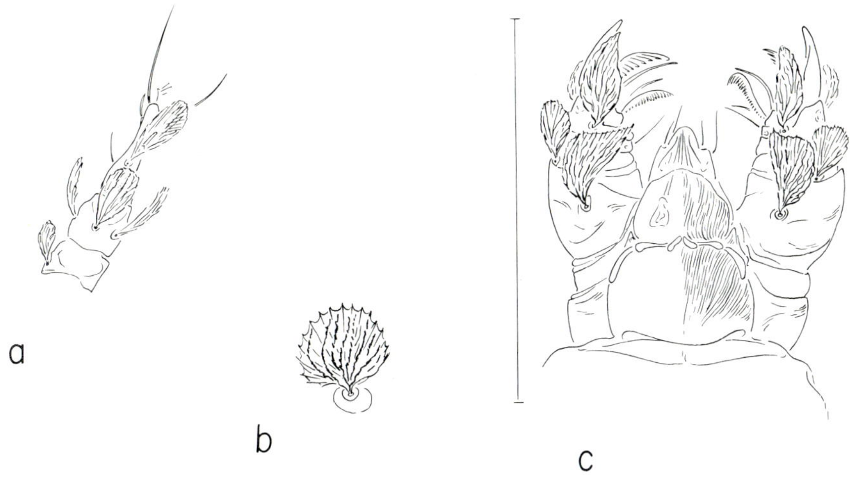

Fig. 43. *Chiapacheylus edentatus* De Leon. *a*, Portion of right leg I; *b*, first dorsolateral propodosomal seta (vertical); *c*, gnathosoma; *d*, female, dorsal aspect.

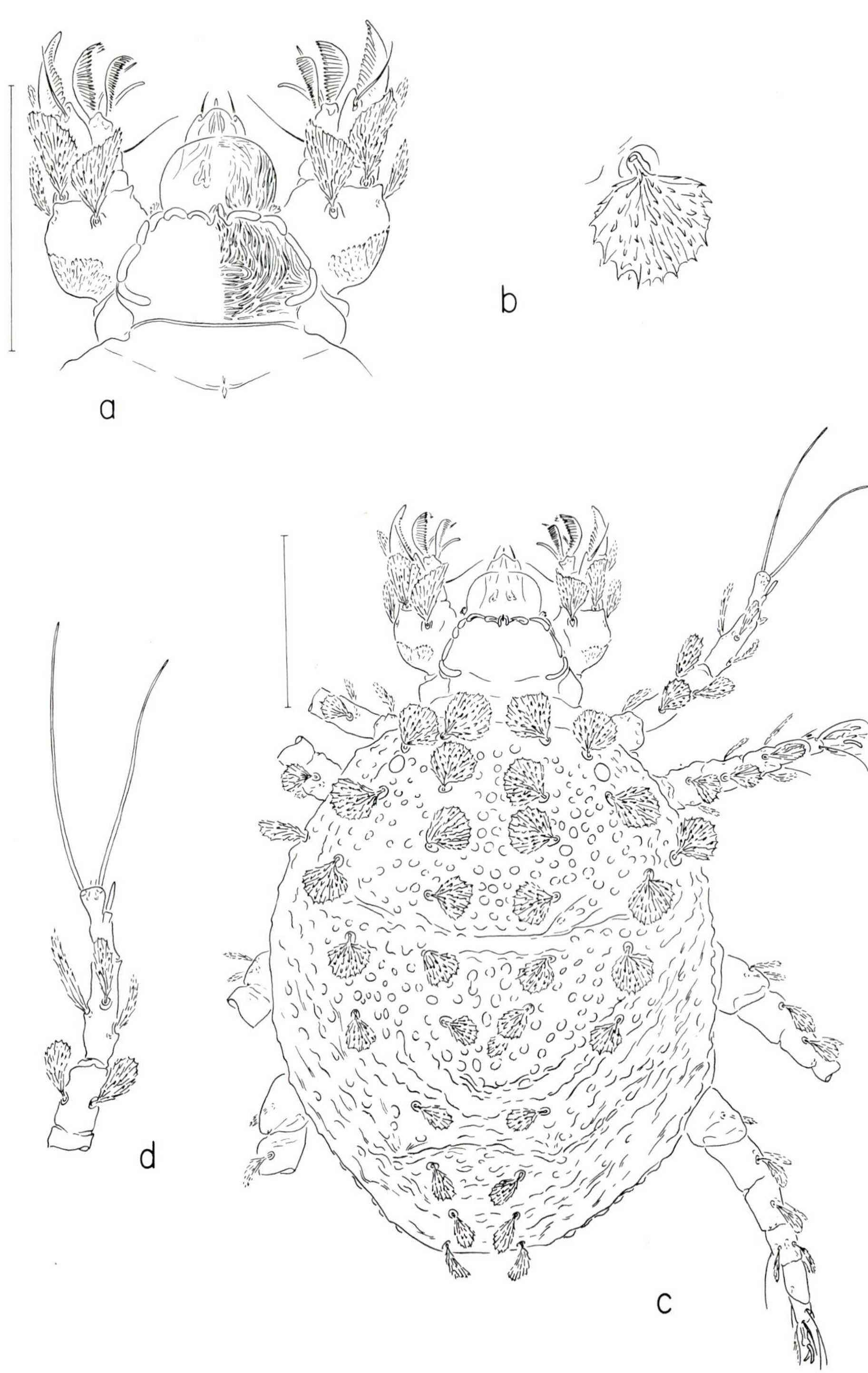

Fig. 44. *Cheletogenes ornatus* (Can. and Fanz.). *a*, Gnathosoma; *b*, first dorsolateral hysterosomal seta; *c*, female, dorsum; *d*, three distal segments of right leg i.

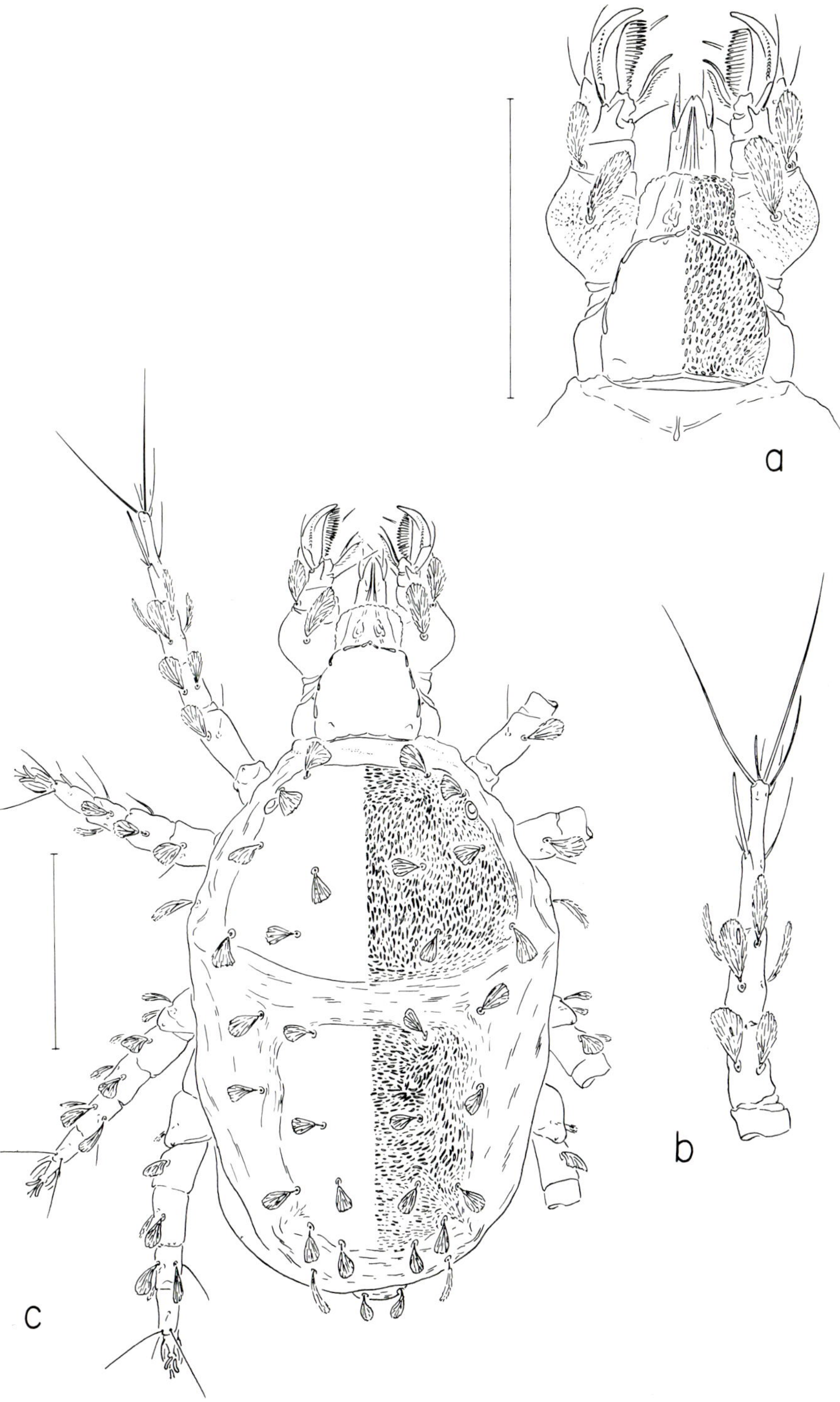

Fig. 45. *Prosocheyla oaklandia* (Baker). *a*, Gnathosoma; *b*, left leg I; *c*, female dorsum.

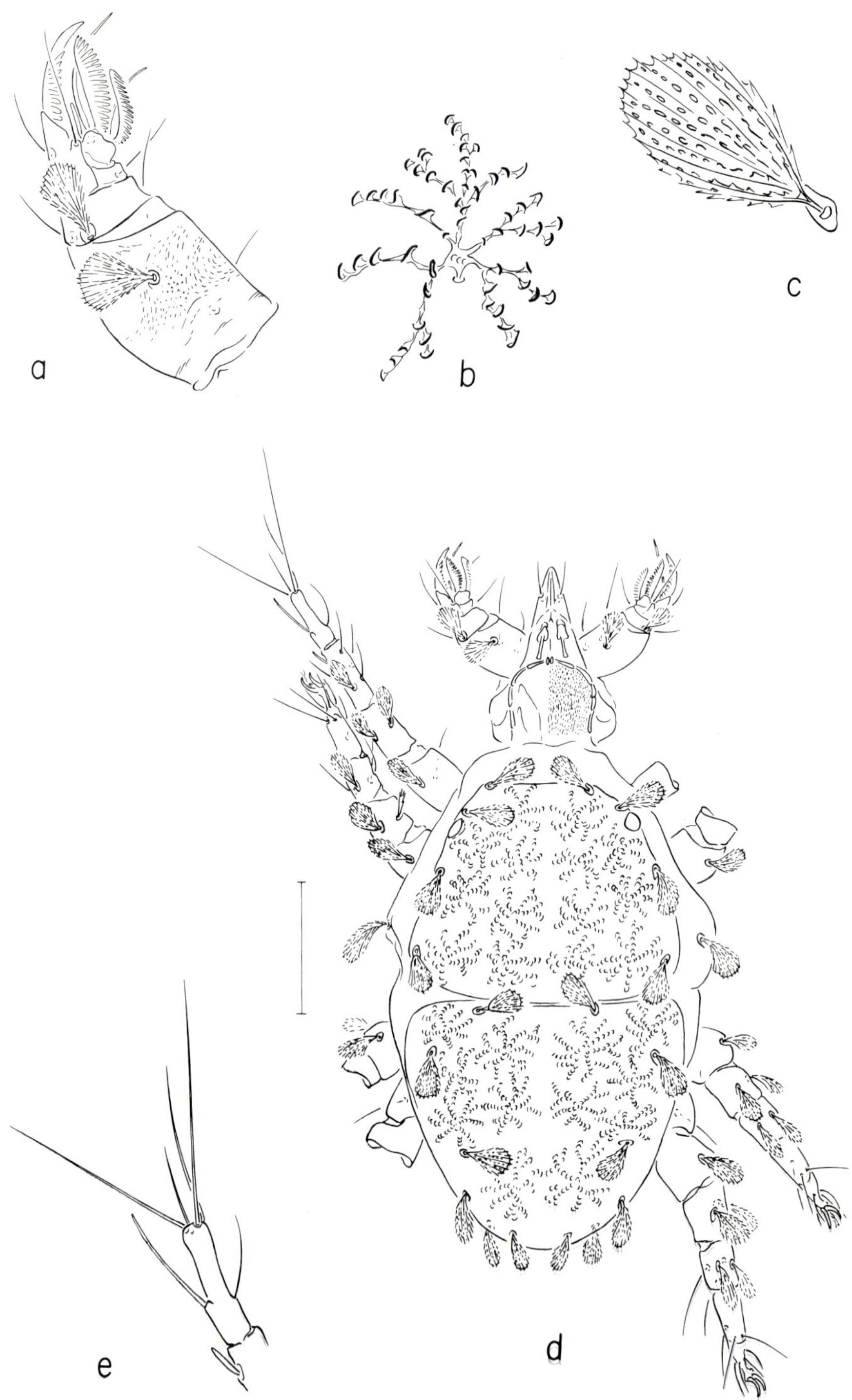

Fig. 46. *Prosocheyla traubi* (Baker). *a*, Left palpus; *b*, a median seta of propodosoma; *c*, first dorsolateral propodosomal seta (vertical); *d*, female, dorsal aspect; *e*, left tarsus I.

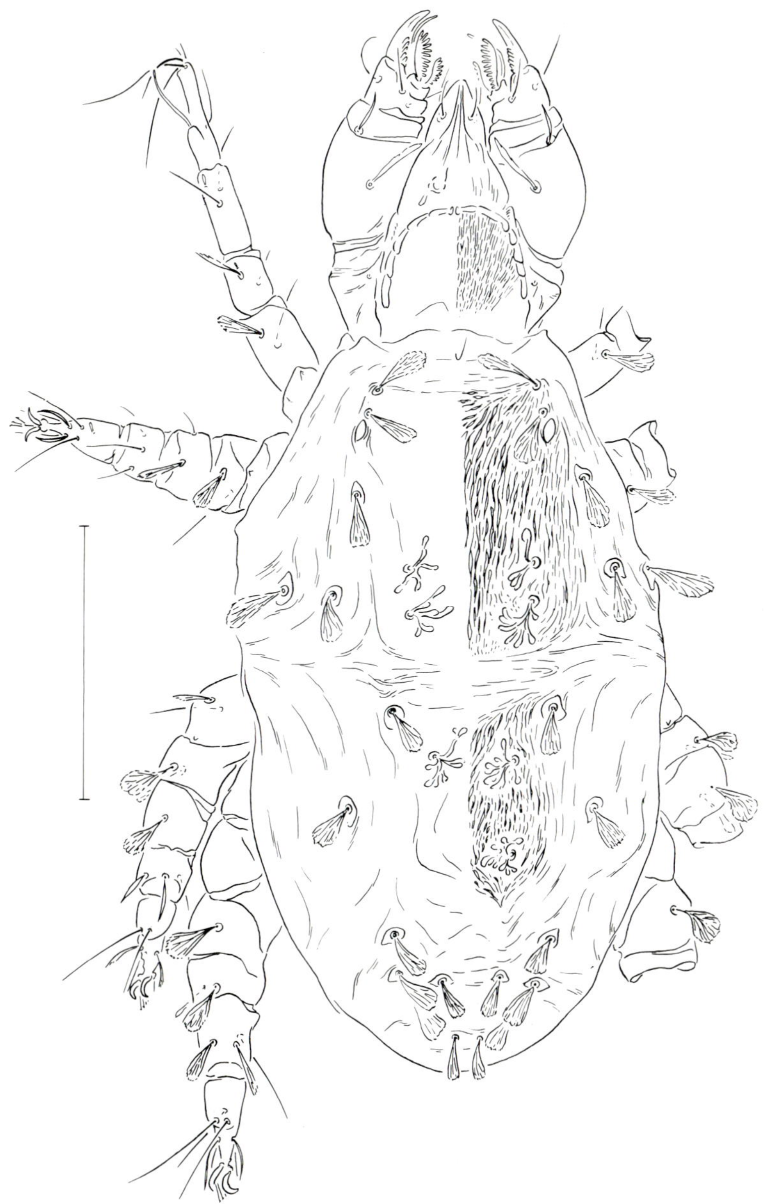

Fig. 47. *Prosocheyla buckneri* (Baker). Dorsal view of female.

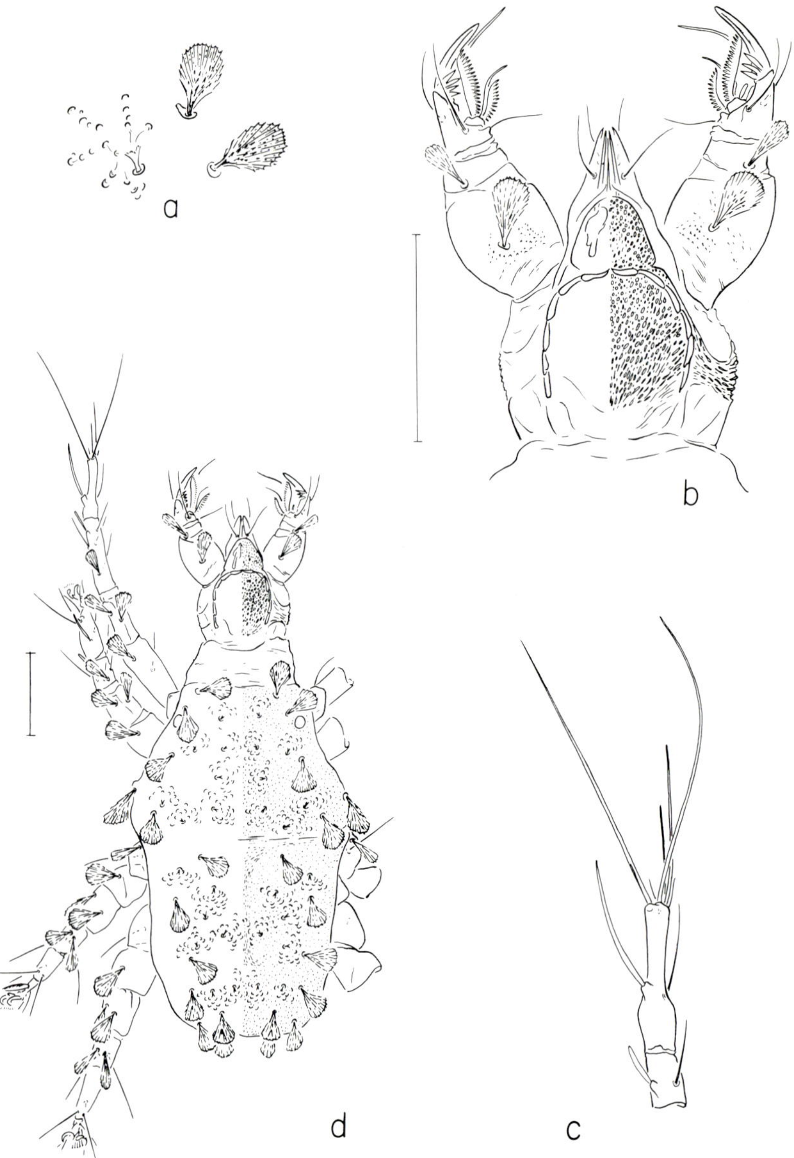

Fig. 48. *Prosocheyla acantha* Smiley and Moser. *a*, First dorsomedian and first two dorsolateral setae on propodosoma; *b*, gnathosoma; *c*, left tarsus I; *d*, dorsal aspect of female.

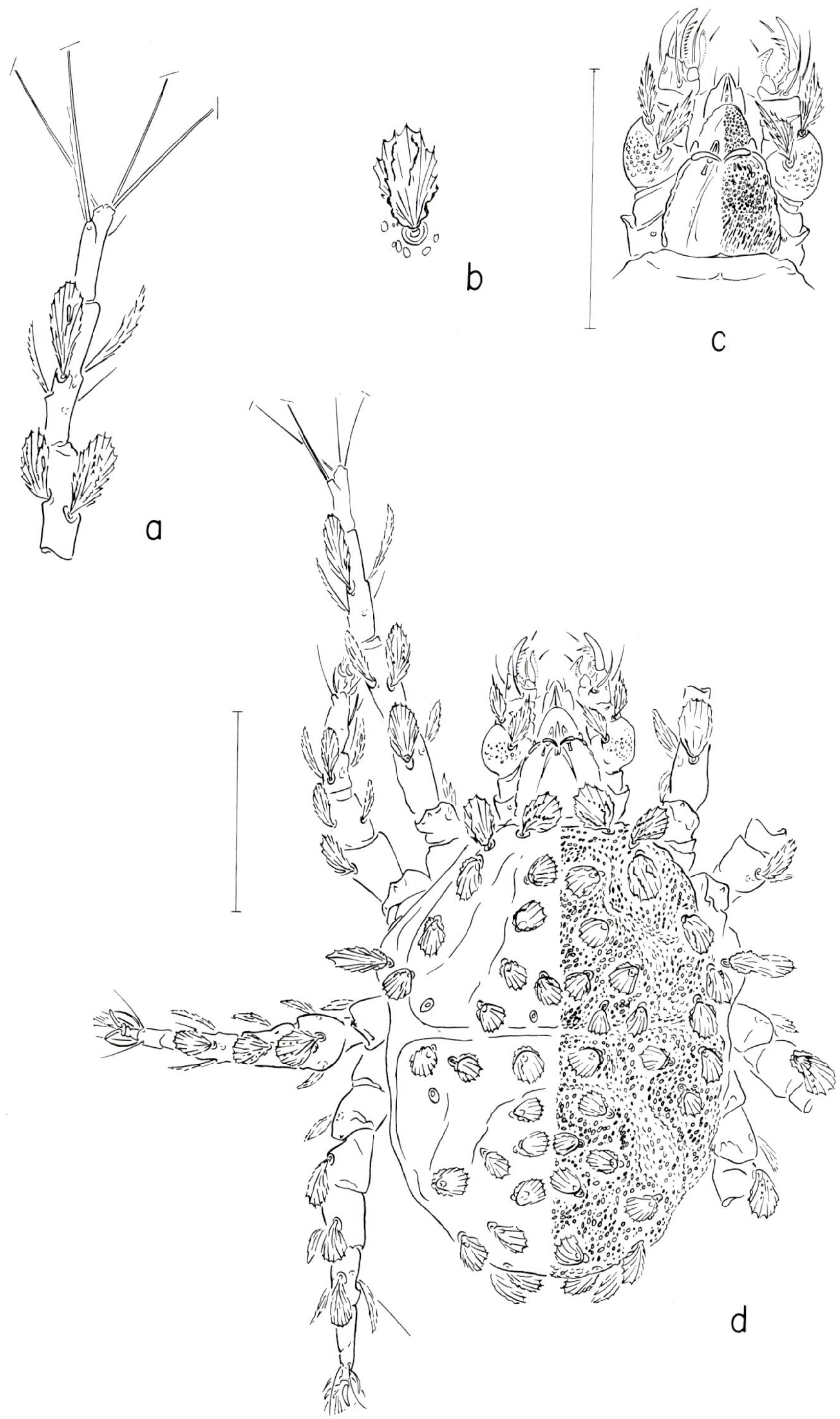

Fig. 49. *Eutogenes foxi* Baker. *a*, Distal segments of left leg I; *b*, second dorsolateral propodosomal (i.e., preocular) seta; *c*, gnathosoma; *d*, female, dorsum.

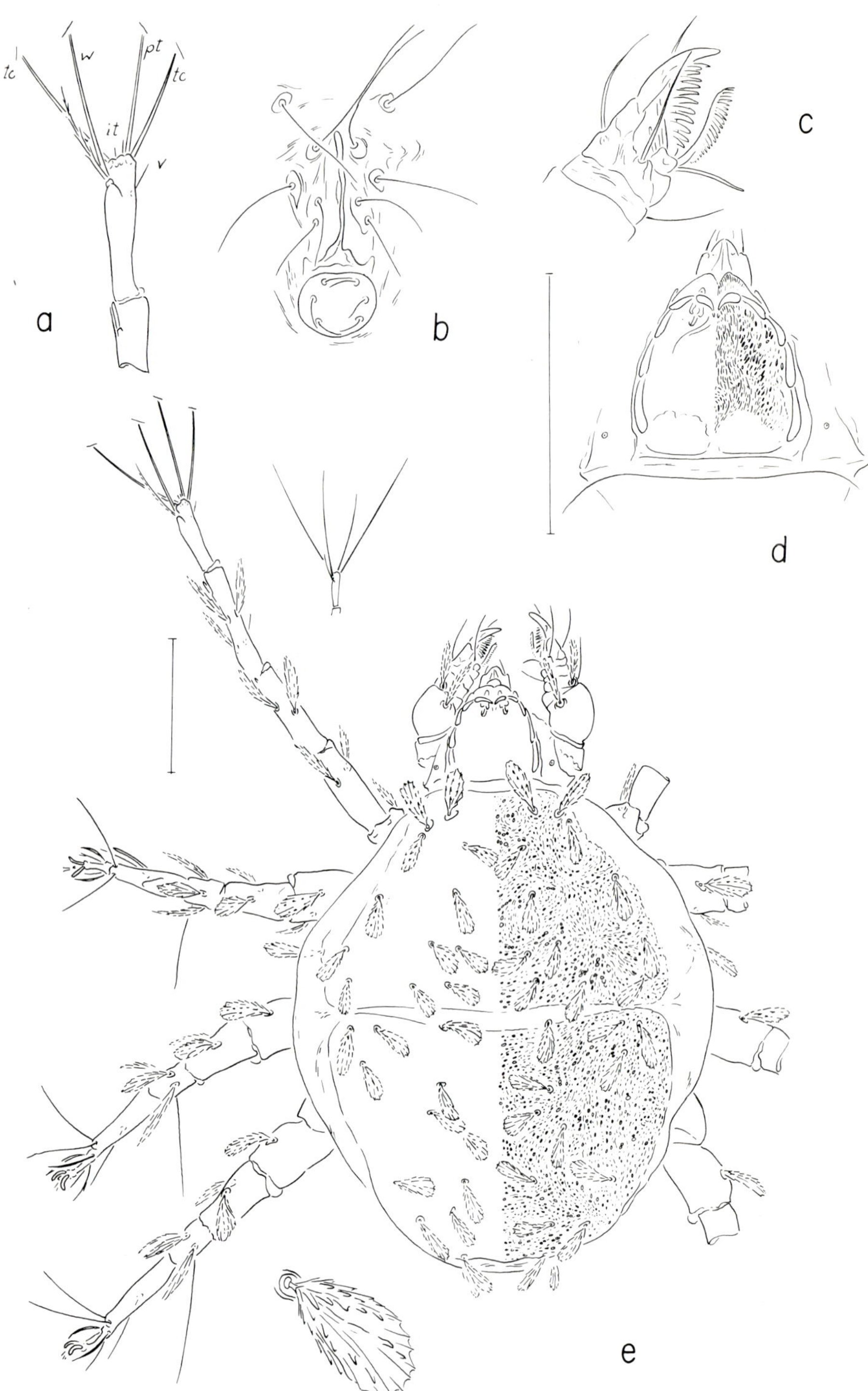

Fig. 50. *Eutogenes vicinus*, n. sp. *a*, Tarsus ɪ; *b*, anogenital region; *c*, tip of right palpus; *d*, stylophore and rostrum; *e*, dorsal aspect of female with first dorsolateral hysterosomal seta drawn below.

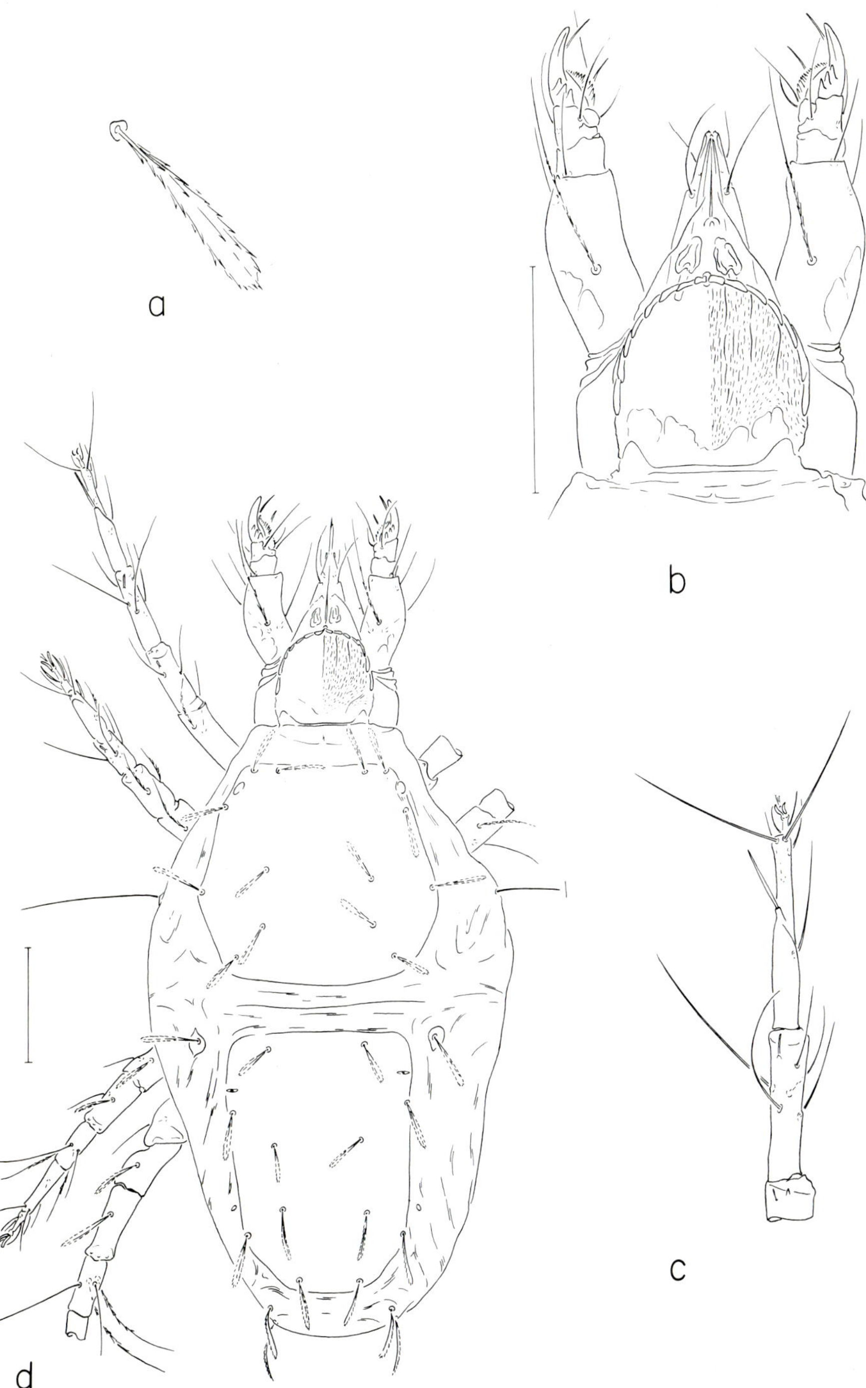

Fig. 51. *Acaropsis sollers* Rohdendorf. *a*, First dorsolateral hysterosomal seta (platelet omitted);
b, gnathosoma; *c*, two segments of left leg I; *d*, female, above.

Fig. 52. *Acaropsella kulagini* (Rohdendorf). *a*, Gnathosoma; *b*, female, dorsum; *c*, left tibia and tarsus I.

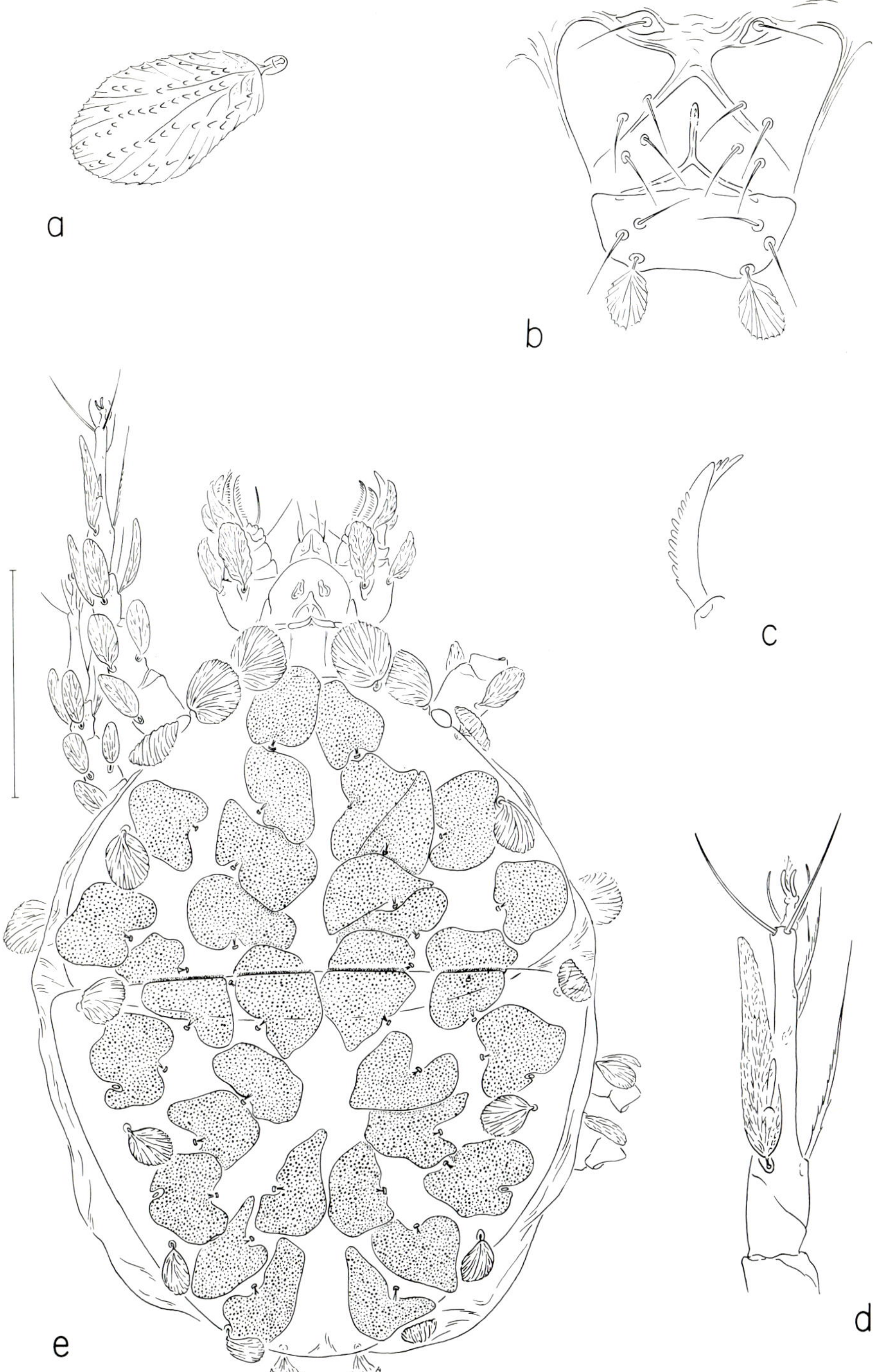

Fig. 53. *Hypopicheyla elongata* Volgin. *a*, Humeral seta; *b*, anogenital region; *c*, right palp claw in lateral view, with part of outer comb projecting from behind; *d*, tarsus I; *e*, dorsal view of female.

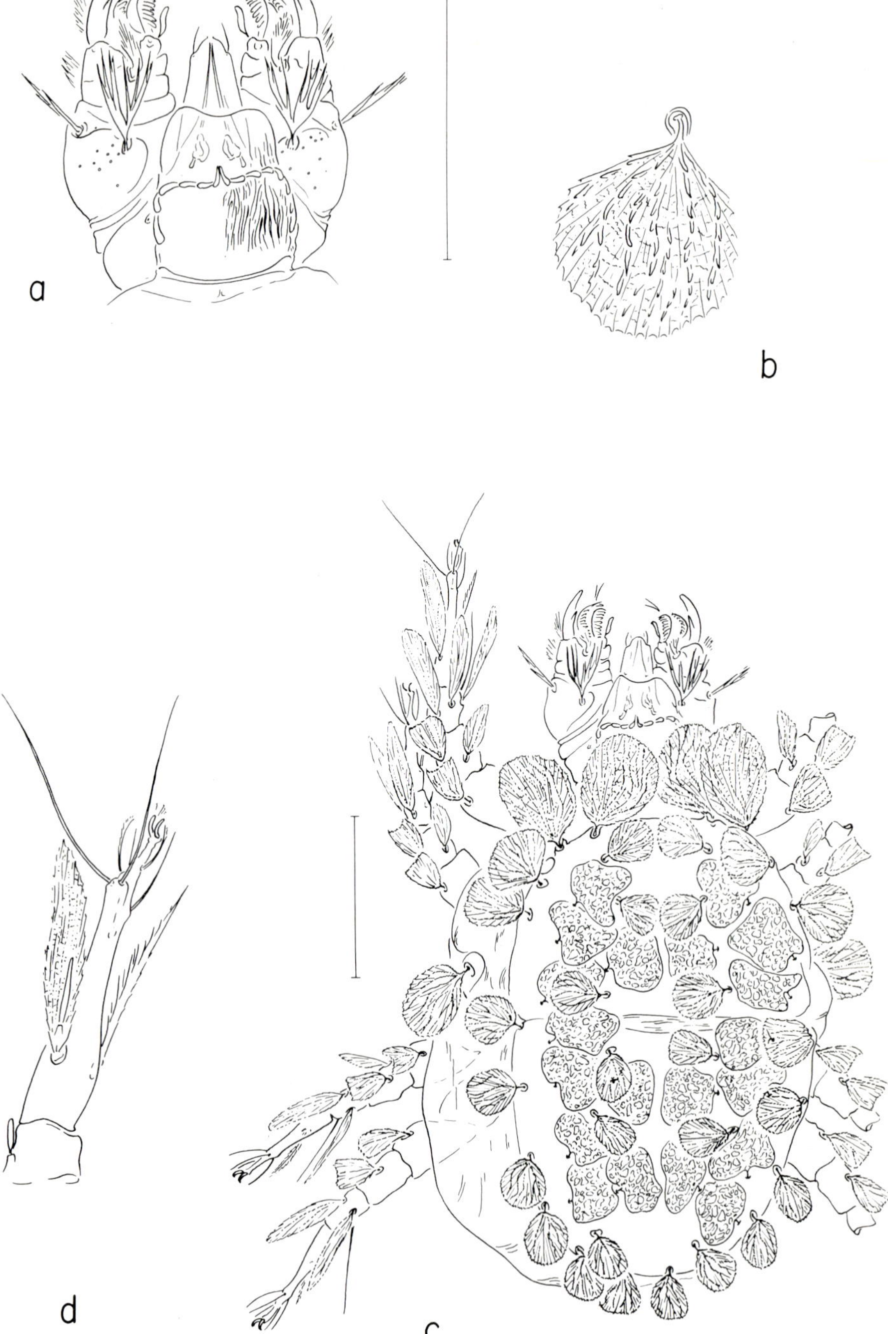

Fig. 54. *Neoeucheyla typhosa*, n. sp. *a*, Gnathosoma; *b*, fourth dorsolateral hysterosomal seta; *c*, female, above; *d*, left tarsus I.

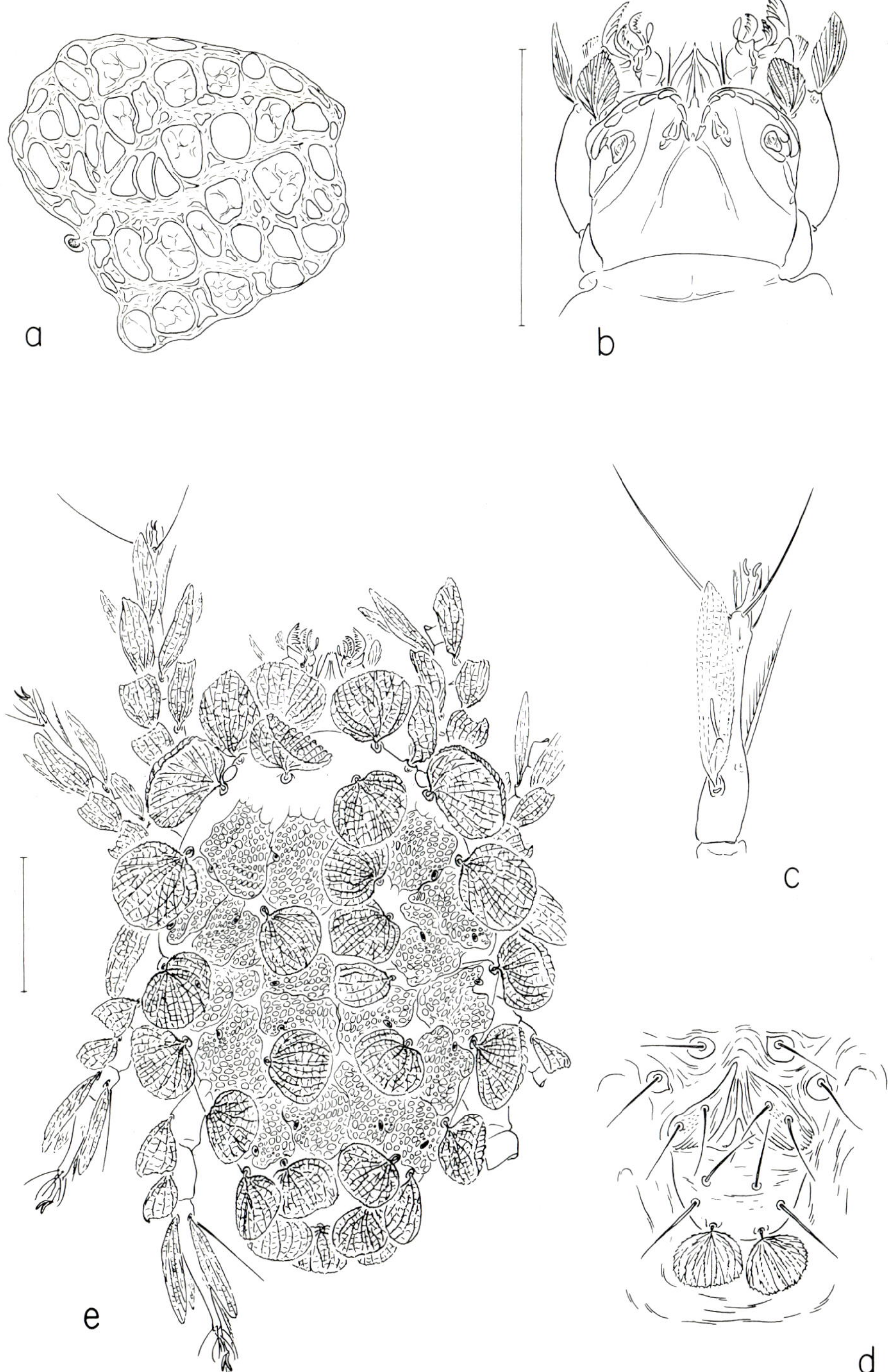

Fig. 55. *Cunliffella panamensis* (Baker). *a*, Squamate dorsomedian seta; *b*, gnathosoma; *c*, left tarsus I; *d*, anogenital region; *e*, female, dorsal aspect.

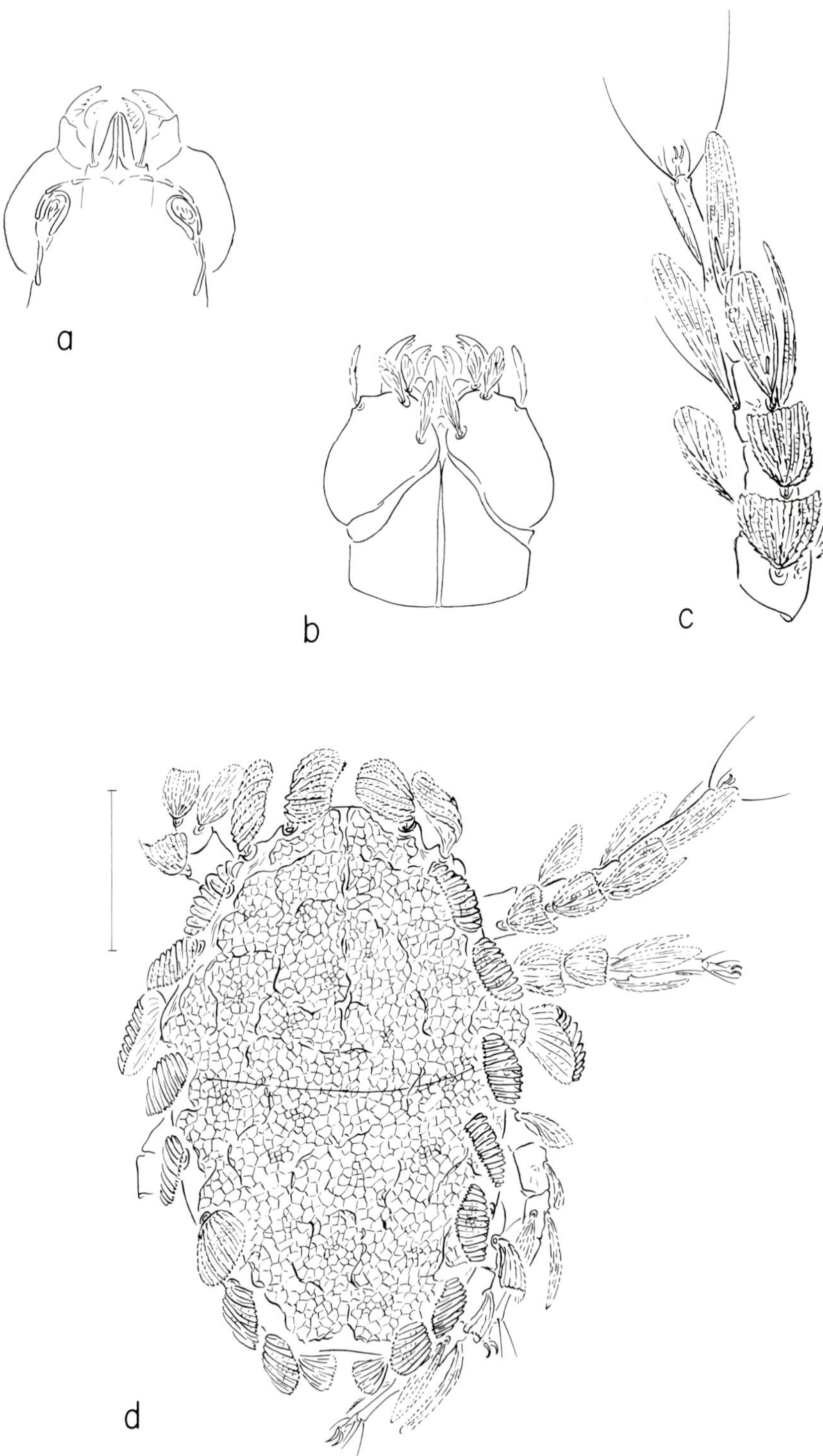

Fig. 56. *Cunliffella whartoni* (Baker). *a*, Dorsal view of gnathosoma as seen through body; *b*, ventral aspect of gnathosoma; *c*, right leg I; *d*, dorsal aspect of female.

Fig. 57. *Microcheyla parvula* Volgin. *a,* One of dorsolateral body setae; *b,* anogenital region; *c,* female, dorsum; *d,* left leg I.

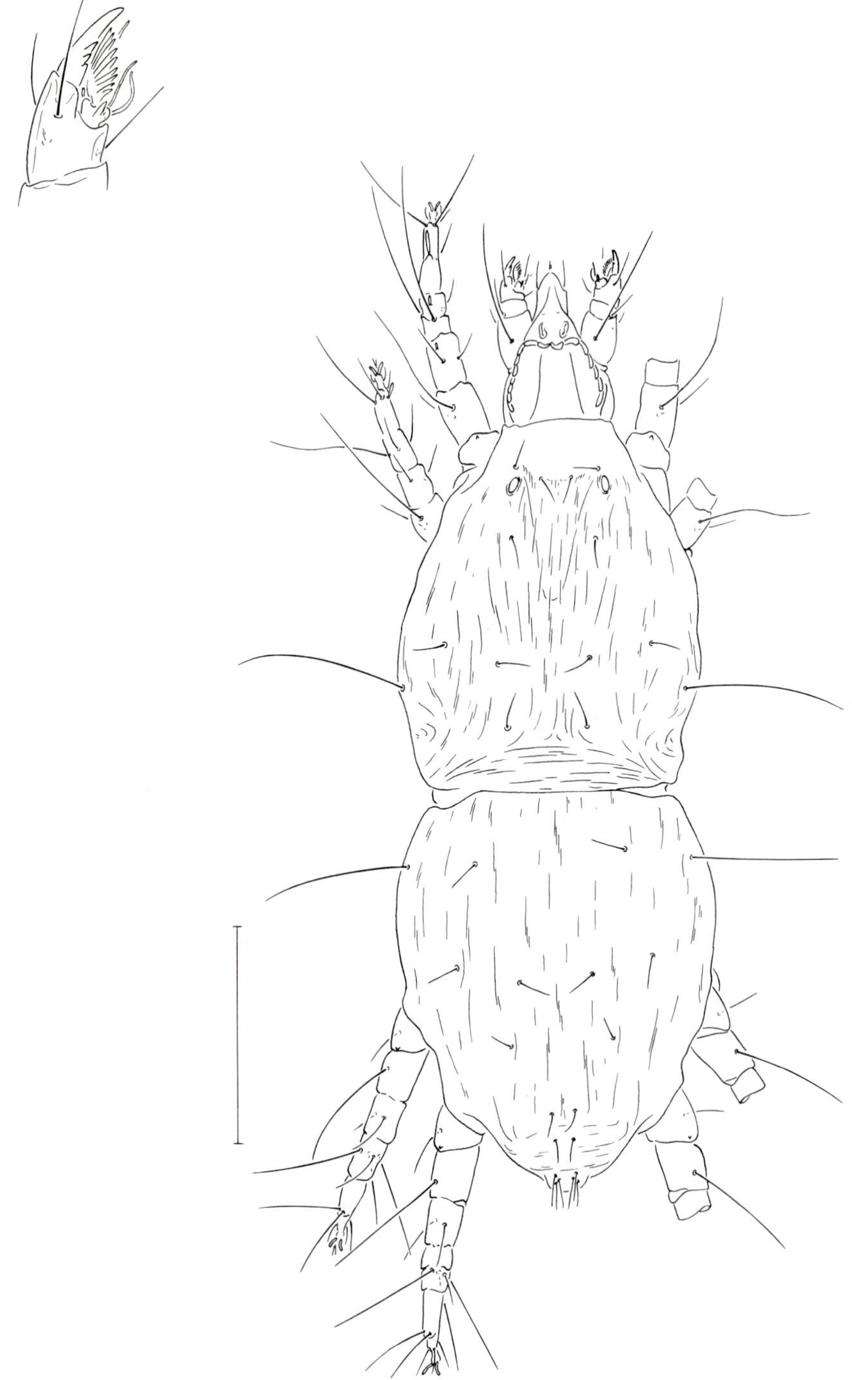

Fig. 58. *Chelacheles bipanus*, n. sp. Tip of left palp shown at upper left.

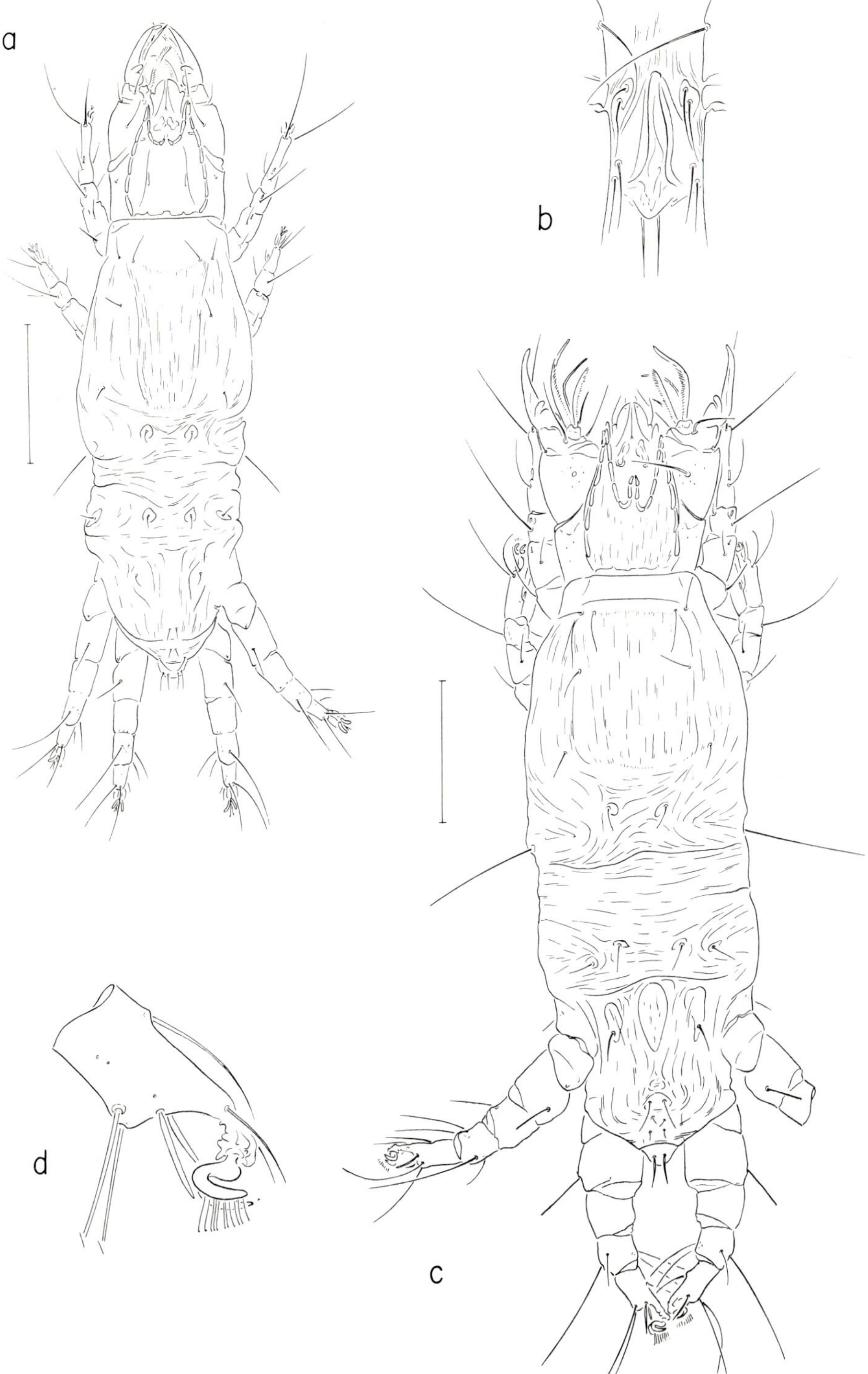

Fig. 59. *Bak micidus*, n. sp. *a*, Dorsal aspect of female. *Bak sanctaehelenae* Yunker. *b*, Anogenital region; *c*, dorsal view of female; *d*, left tarsus IV.